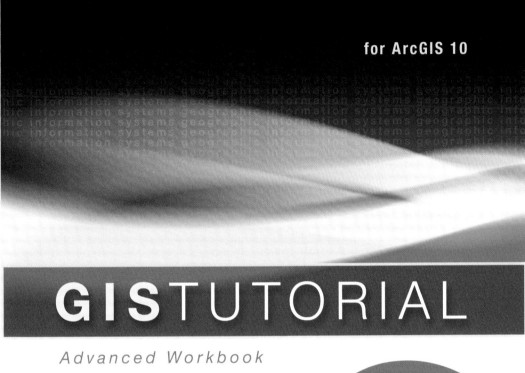

for ArcGIS 10

GIS TUTORIAL

Advanced Workbook

David W. Allen
Jeffery M. Coffey

ESRI PRESS
REDLANDS, CALIFORNIA

Esri Press, 380 New York Street, Redlands, California 92373-8100

15 14 13 12 11   2 3 4 5 6 7 8 9 10 11

Printed in the United States of America

Library of Congress Cataloging-in-Publication Data
Allen, David W., 1961-
    GIS tutorial 3 : advanced workbook / David W. Allen, Jeffery M. Coffey.
      p. cm.
    Includes index.
    "For ArcGIS 10."
    ISBN 978-1-58948-207-4 (pbk. : alk. paper)
    1.   Geographic information systems. 2.   Spatial analysis (Statistics) 3.   ArcGIS. I. Coffey, Jeffery M. (Jeffery Morgan) II. Title. III. Title:
Geographic information systems tutorial three.
    G70.212.A435 2010
    910.285--dc22   2010012696

Ask for Esri Press titles at your local bookstore or order by calling 800-447-9778, or shop online at www.esri.com/esripress. Outside the United States, contact your local Esri distributor or shop online at www.eurospanbookstore.com/Esri.
Esri Press titles are distributed to the trade by the following:

In North America:
Ingram Publisher Services
Toll-free telephone: 800-648-3104
Toll-free fax: 800-838-1149
E-mail: customerservice@ingrampublisherservices.com

In the United Kingdom, Europe, Middle East and Africa, Asia, and Australia:
Eurospan Group
3 Henrietta Street
London WC2E 8LU
United Kingdom
Telephone: 44(0) 1767 604972
Fax: 44(0) 1767 601640
E-mail: eurospan@turpin-distribution.com

Contents

Preface

The study of geographic information systems (GIS) is a wide-ranging topic, and ArcGIS software covers an extensive technology. There are many books written about how the software works and how it can be used to manipulate data and solve spatial problems. So, the task becomes defining where this book fits into your educational cycle of learning all you can about GIS.

GIS Tutorial 3: Advanced Workbook concentrates mainly on features of the ArcInfo and ArcEditor licenses of ArcGIS that are not available at lower license levels. While it may seem on the surface to jump from topic to topic, remember that it is not trying to demonstrate basic software features—there are lots of books that do that. It is instead designed to instruct you on the use of the more complex functions of the software. Incorporating these into your daily workflow will help make your job a little easier and your maps a little better. You should be familiar with the skill sets learned in *GIS Tutorial 1* and *GIS Tutorial 2* and have experience working with shapefiles, feature datasets, and feature classes in ArcMap.

GIS Tutorial 3: Advanced Workbook contains nine chapters, each containing tutorials and exercises that build pertinent skills. A task index is included to help you locate relevant ArcGIS tasks and tools within the book. Instructors will find data, a course syllabus, and teacher guides on the instructor DVD available upon request from www.esri.com/esripress. All 23 tutorials and related exercises can be covered in a fast-paced 16-week semester. Instructors may also select chapters that are most relevant to their program and complete those chapters at a slower pace during the semester.

The first part of the book, containing three chapters, explores the steps in designing schemas and creating data from scratch. Many other books start with premade datasets and instruct you on how to manipulate and expand the data to accomplish something, but where does this data come from? Predesigned database schemas are also available that can be used as templates for your datasets. While these can be a great starting point, there will ultimately be changes that you will want to make. Only by fully understanding the process of designing database schemas will you be able to make changes that ensure the best and most efficient use of the geodatabase. There are also situations where a premade data schema is not available for the geodatabase you need. Has anyone, for example, done the definitive design on routing a street sweeper and recording its progress? If given this project, you would have to design and create the schema yourself. In chapter 1, you will learn how to design a functional, efficient geodatabase, and in chapter 2, you will create it. Chapter 3 shows how to bring existing data into a complex geodatabase that you create.

Part 2, which contains two chapters, deals with creating new features in the ArcMap environment. The editing and topology tools in chapters 4 and 5 can be used to take advantage of the geodatabase design and make the best use of the data integrity rules. There are also many techniques shown to speed up the creation process, sometimes to the point of having the software create features for you. By using these, the resulting data not only will be more accurate, but may also be easier to maintain in the future.

Part 3, containing two chapters, deals with managing workflows in ArcGIS. The use of custom tools, scripts, and custom toolbars explored in chapter 6 can make you much more successful in your editing tasks. Even the simplest customization might save several keystrokes or mouse clicks, resulting in a time savings. More complex techniques, such as scripts written with the Python programming language, will have a bigger impact on your workflow. Chapter 7 shows how to automate your workflow in a visual scripting environment called ModelBuilder, where you can develop custom tools with the look and feel of standard ArcGIS tools.

Part 4, containing two chapters, is aimed at helping you use the more advanced cartography tools available in ArcGIS. Chapter 8 deals with advanced labeling techniques and annotation, while chapter 9 concludes with ways to customize your legend and make the most of cartographic representations to draw advanced symbology. The end result will be a map that is more engaging—one that presents your ideas and analysis clearly and concisely.

The set of skills covered in this book should take you beyond the basics and introduce you to tools and methods that use the best features of the higher license levels of ArcGIS.

This book comes with a DVD containing exercise and assignment data, and a DVD containing a trial version of ArcGIS Desktop 10. You will need to install the software and data in order to perform the exercises and assignments in this book. (If you have an earlier version of ArcView, ArcEditor, or ArcInfo installed, you will need to uninstall it.) The ArcGIS Desktop 10 DVD provided with this book will work for instructors and basic-level students in exercise labs that previously used an ArcView license of ArcGIS Desktop. Instructions for installing the data and software that come with this book are included in appendix D.

For teacher resources and updates related to this book, go to www.esri.com/esripress.

Acknowledgments

David W. Allen: I get a lot of students in my GIS classes at Tarrant County College in Arlington, Texas, who have been working in the GIS industry for years and have picked up GIS skills along the way. These students have greatly benefited from structured training on some of the more advanced tools in ArcEditor and ArcInfo, which prompted me to write this book. Jeff Coffey of the Tarrant Regional Water District in Fort Worth came on board to help polish this into a complete package, and Judy Hawkins of ESRI Press helped us get the manuscript noticed. A debt of gratitude goes out to each of them, as well as to all the students who have suffered through my classes over the years. It was their puzzled looks and questions that prompted me to rewrite sections of the manuscript to offer better explanations and clearer instructions.

I'd also like to thank Carolyn Schatz of ESRI Press, who edited this manuscript. Although she is not a GIS expert by trade, she would probably rival most of you after having worked through these tutorials so many times. It's a tough process, but she made it very enjoyable. Other great people at ESRI Press include editorial supervisor David Boyles; cartographer Riley Peake, who did the technical reviews; and the rest of the ESRI Press production, marketing and distribution staff.

And finally, thanks to the City of Euless administration, which allowed its rich GIS datasets to be used in the making of these tutorials. While the data and processes are based on reality, all the scenarios are fictional and should not be associated with the City of Euless.

Jeffery M. Coffey: Having worked in the public, private, and academic sectors utilizing GIS for 20-plus years, I immediately recognized the benefits of developing a textbook aimed at students already well acquainted with GIS technology but who seek a deeper understanding of some of the advanced applications available with ArcGIS. I feel confident that this text will give students additional insight and skills to carry the application and use of GIS technology into the future. I would like to thank David Allen for offering me the opportunity to help with this book, the excellent editorial staff support at ESRI Press, my family at the Tarrant Regional Water District, and my girls at home—Robin, Morgan, and Ella—for their patience and support.

Part 1

**Designing a framework
for the complex geodatabase**

Designing the
geodatabase schema

Tutorial 1-1

Creating a geodatabase—building a logical model

The past few releases of ArcGIS software have included many additions and improvements to the data storage capabilities of the geodatabase file structure. There are new techniques to control interaction with the data, assign a behavior to it, and define relationships among datasets. In the design process, it is important to understand these techniques in order to build the most efficient database possible.

Learning objectives

- *Outline geodatabase behavior*
- *Integrate datasets*
- *Model reality*
- *Use ArcCatalog*

Introduction

The goal of designing a geodatabase is to model the reality it is intended to represent. Many characteristics, or behaviors, of the data can be included in a geodatabase using various techniques. As the data modeler, it is your job to explore the capabilities of ArcGIS to make the most efficient and flexible database possible. The time spent at the start of a project in designing the geodatabase will reap rewards later by making the data easier to use and edit and by presenting a better representation of reality.

The first step is to study the reality that is about to be modeled. Look carefully and determine what features will need to be included in the geodatabase. In ArcGIS, everything is modeled as points, lines, and polygons, so realistic characteristics will need to be assigned to these pieces.

Next, look at how the data will be created. Will it be imported from another source, collected with field equipment, traced from aerial photos, drawn from survey data, or be derived from some other process?

1–1

1–2

Finally, consider how this data will be used. Who will do the edits, and what queries might it be expected to support in the future? Knowing what will be asked of the data later will have a great impact on what the design will look like now.

With these things in mind, you can start to construct a logical data model. The model will diagram your process outline and allow for updates and changes before the final design is committed to a geodatabase. There are many tools available to diagram geodatabases, including the Geodatabase Diagrammer available by download from the ArcScripts Web site (`http://arcscripts.esri.com/`), and available on the included DVD. This, however, requires the purchase of additional software to run, so as an alternative, there are Microsoft Excel spreadsheet files included on the DVD that may also be used for basic diagramming.

The logical model is used to show what data types you will have (points, lines, or polygons), what tabular data will be included in the dataset, and the relationships between the tabular data and feature classes (if any). The model is also easily shared among your colleagues, so you can get several opinions on the design you are attempting.

Once a preliminary logical model is done, it can be checked against many of the advanced features of the geodatabase such as domains, subtypes, and relationships. Have the best tools been employed to ensure data integrity, ease of editing, and future expansion? You will also check the geodatabase against your idea of how the data should behave.

The result should be a well-thought-out geodatabase that is both efficient and a good representation of reality—well, at least as close as you can get by using points, lines, and polygons.

You will start with a simple geodatabase in this tutorial, and then examine several advanced geodatabase options to see if better efficiency and more realistic behaviors can be achieved. The good news about any geodatabase design is that if it works, it's a success. Ten different people could design ten different geodatabases for the same project, and they all could work quite well. The true test is how efficient a geodatabase is, how well it models reality, how well it maintains data integrity, and how easy it is to work with in editing and extracting information.

Designing the data

Scenario   The City of Oleander has hired you as the top-gun geodatabase designer and wants an all new database design for its parcel data. The new geodatabase will be used to designate each piece of property in the city—who owns it, its legal description, address, and much more. You'll get information from the city planner to get the full idea of what's needed and then create a diagram of your proposal. At this point, the geodatabase will not be constructed, only designed.

The city planner describes a dataset that would have a polygon for each piece of property in the city, whether it is platted or unplatted. It should have information about the legal description, the street address, and the current usage.

Data Since you are creating this geodatabase from scratch, there is no data to start with. But you will need to print the geodatabase design forms on the DVD as an aid in the design process.

Tools used Geodatabase design forms

Begin the logical design for the geodatabase

The main component of this geodatabase will be polygons representing every piece of property in the city. Each piece of property is assigned certain data by the city. This includes the subdivision name, block designation, lot designation, street address, and a land-use code, which shows how the land is being used.

As you know, the geodatabase is the framework in which other components are built. It may contain feature classes, tables, relationship classes, feature datasets, and many other components. You will design the geodatabase and its components using the geodatabase design forms provided on the accompanying DVD.

1 Open the file C:\ESRIPress\GIST3\Data\GDB Design Forms in either the Excel or PDF format. Print all six pages of the geodatabase design forms (GDB feature classes, tables, domains, domains2, subtypes, and relationships). The first two pages (feature classes and tables) will be used for these first few steps, but the other pages will be used in the other steps of this tutorial.

The geodatabase will need a name. This should reflect in general terms what will be stored in it.

2 On the first line of the first page (GDB feature classes), write the name **LandRecords** for the geodatabase name. The next line asks for a feature dataset name. A feature dataset is used to separate data into smaller subsets but is also important in grouping data for use in topology and various other advanced features. For now, leave the feature dataset name blank.

The next step is to start filling in the feature classes. So far, the city planner has only described one feature class. This will contain the polygons representing parcels and include the fields the city planner described.

3 On the feature classes worksheet, add a new feature class named **Parcels**. Note its type as **POLY** (for polygon) and give it an alias of **Property Ownership**.

Geodatabase name	LandRecords
Feature dataset name	

Feature classes:

Type	Feature class name	Alias
POLY	Parcels	Property Ownership

Type:	Indicate if this is a **PoiNT**, **L**ine, or **POLY**gon feature class.
Name:	Enter the name of your feature class.
Alias:	Describe the contents of the feature class.

1-1

1-2

The alias is one of the first characteristics of the geodatabase that will be assigned. This alias will be shown in the table of contents when the layer is added to ArcMap, and consequently, it also could be used in a legend. It should be very descriptive of the data to distinguish it from other datasets.

This feature class will have fields to store data, and these are recorded on the second design form. From the city planner's description, you can determine that the feature class will have fields for the subdivision name, block designation, lot designation, street address, and a land-use code. The first three fields are pretty straightforward. They will need to contain alphanumeric characters, so their field types will be Text.

4 On the tables worksheet, write the name of the new feature class on the first line. Under field name, enter the first field as **SubName**. Note its type as **Text**. Add another field for **Blk** as **Text** and **LotNo** as **Text**.

Tables worksheet

Feature class or table name	Field name	Field type	Alias	Nulls (Y/N)	Default value	Domain name or subtype field (D) or (S)
Parcels	SubName	Text				
	Blk	Text				
	LotNo	Text				

Pretty simple so far, but there is other information to enter that will affect the future use of the data. The first is the field alias, which is another characteristic of the geodatabase. This alias will be shown in many of the ArcGIS tools, the attribute table, any classification schemes, and many more places when the data is accessed. The field alias should describe what data the field contains.

5 Next to the field name SubName, write the description **Subdivision Name** as the alias. Then write the alias **Block Designation** for Blk and **Lot Number** for LotNo.

Tables worksheet

Feature class or table name	Field name	Field type	Alias	Nulls (Y/N)	Default value	Domain name or subtype field (D) or (S)
Parcels	SubName	Text	Subdivision Name			
	Blk	Text	Block Designation			
	LotNo	Text	Lot Number			

That's taken care of some of the fields. The next is street address information. This could be entered as a single field, but if you ever want to geocode against this dataset, it would be better to have each component of the address in separate fields. The common fields for geocoding are street prefix type, prefix direction, address number, street name, street type, suffix direction, and ZIP Code. Fields such as city name or state name may be necessary if you are geocoding a broader region, but since all the features that this dataset will contain are specific to Oleander, you can leave them out. All the listed fields need to be included in the table.

One interesting thing is that if the fields are given certain names that ArcMap recognizes, they will be filled in automatically when you make an address locator. An address locator is a special file that ArcGIS builds using your dataset that will allow addresses to be found easily when geocoding or using the Find tool. This can also be used in routing and network applications. A list of preferred field names for each field is stored in the address locator style file. You can then open the file in the \ArcGIS\Locators folder and view the list by searching for the phrase Preferred Field Names. You may add your own field names to the list as needed.

6 Open a Windows Explorer window. Navigate to the folder containing your ArcGIS installation (e.g., C:\Program Files\ArcGIS) and open the Locators folder. Scroll down to the file USAddress.lot, right-click, and select Open. If given the option, select "Choose the program to open this file," and then select Notepad. Use the Find tool or scroll down to the area labeled House Number.

```
USAddress.lot - Notepad
File  Edit  Format  View  Help
        <field_role name="Primary.House" required="true">
          <display_name>House Number</display_name>
          <preferred_name>HOUSE</preferred_name>
          <preferred_name>HOUSENUM</preferred_name>
          <preferred_name>HOUSE_NUM</preferred_name>
          <preferred_name>HOUSE-NUM</preferred_name>
          <preferred_name>HOUSE_NUMBER</preferred_name>
          <preferred_name>HOUSE-NUMBER</preferred_name>
          <preferred_name>HOUSE_NUMB</preferred_name>
          <preferred_name>HOUSE-NUMB</preferred_name>
          <preferred_name>NUMBER</preferred_name>
          <preferred_name>ADDNUMBER</preferred_name>
          <preferred_name>ADD_NUMBER</preferred_name>
          <preferred_name>ADD</preferred_name>
          <preferred_name>ADDRESS</preferred_name>
          <preferred_name>ADDR</preferred_name>
        </field_role>
        <field_role name="Primary.PreDir">
          <display_name>Prefix Direction</display_name>
          <preferred_name>ADDR_PD</preferred_name>
          <preferred_name>PREFIX</preferred_name>
          <preferred_name>FDPRE</preferred_name>
          <preferred_name>FEDIRP</preferred_name>
          <preferred_name>PREDIR</preferred_name>
          <preferred_name>PRE_DIR</preferred_name>
          <preferred_name>DIR</preferred_name>
          <preferred_name>DIRPABV_AL</preferred_name>
```

As you can see, there are ten or more acceptable field names for each of the fields necessary for address geocoding. Look over the list to see what the choices are.

7 On the tables worksheet, write the following field names, field types, and aliases:

- Pre_Type, Text, Prefix Type
- Pre_Dir, Text, Prefix Direction
- House_Num, Text, House Number
- Street_Name, Text, Street Name
- Street_Type, Text, Street Type
- Suffix_Dir, Text, Suffix Direction
- ZIPCODE, LI (Long Integer), ZIP Code

These fields will add a lot of functionality to the dataset that may be valuable later. You could select all the parcels in a certain subdivision, all the parcels that front a certain street, or use the House_Num field to put address labels on the map.

The last bit of data that the city planner mentioned was the land-use code. This is a two- to nine-letter alphanumeric code denoting how the land is currently being used. Add this to the database model.

8 On the tables worksheet, add the field **UseCode** with the field type **Text** and the alias **Land Use Code**. The data entered so far has had to do with information that the city planner wanted. One more piece of data will be necessary for you to maintain a connection to certain third-party data that is important to the project. The identity of the property owner is not stored in the parcel's attribute table but is stored in an external table. You will need to add a field to your data structure that will allow you to set up a relationship between it and the external table. The procedure is discussed later, but for now, you will need to add a field to accommodate the relationship. The field should be called Georeference and have a field type of Text.

9 On the tables worksheet, add the field **Georeference**, with the field type **Text** and the name **Georeference**.

Design for data integrity

The design looks pretty good, but imagine what will happen when people start putting data in the table. If they were to leave the SubName field blank, there would be no way to identify the legal record of a piece of property. What about address number or land-use code? Leaving these blank could create gaps in the data. On the other hand, not every street will have a value for prefix type, so there will be instances where a space can be left blank and still be correct.

One way to build data integrity rules into your table is to set the flag for allowing nulls, or no value, for a field. If nulls are not allowed, ArcMap will produce an error for any records entered without all the necessary values being provided. Perhaps, the person entering the data accidentally skips the field during data entry or tries to enter data before all the information is known. Either way, it could cause problems with your data.

The solution is to mark in the design table which fields are allowed to have nulls and which must have a value entered.

1 On the tables worksheet, mark the following fields to allow null values:

- Pre_Type
- Pre_Dir
- Suffix_Dir

2 **Mark the remaining fields as not allowing null values.** Another data integrity component is the domain. A domain allows you to define a list of values for any text field or a range of values for a numeric field. When data is entered, it is matched against the domain to see if it is a valid value. This helps eliminate typos or inventive abbreviations. Imagine ten data entry clerks all coming up with unique abbreviations for the land-use code Vacant. It might be entered as VAC, V, Vcnt, or any number of misspellings. A query to find all vacant property would be difficult. If a domain is applied to the field Use_Code with all the correct abbreviations, it would be impossible for anyone to enter a value that wasn't in the domain.

The domain values will be entered on the domains worksheet, and you will note that it is a domain to avoid confusion with subtype fields that may be entered later.

1–1
1–2

3   On the tables worksheet, note on the line for UseCode the name of a domain that will contain the acceptable values for this field. Call it **ParcelUseCodes** and place a **(D)** in front of it for domain.

Tables worksheet Feature class or table name	Field name	Field type	Alias	Nulls (Y/N)	Default value	Domain name or subtype field (D) or (S)
Parcels	SubName	Text	Subdivision Name	No		
	Blk	Text	Block Designation	No		
	LotNo	Text	Lot Number	No		
	Pre_Type	Text	Prefix Type	Yes		
	Pre_Dir	Text	Prefix Direction	Yes		
	House_Num	Text	House Number	No		
	Street_Name	Text	Street Name	No		
	Street_Type	Text	Street Type	No		
	Suffix_Dir	Text	Suffix Direction	Yes		
	ZIPCODE	LI	ZIP Code	No		
	UseCode	Text	Land Use Code	No		(D) ParcelUseCodes
	Georeference	Text	Georeference	No		

4   Now turn to the domains worksheet and write the domain name **ParcelUseCodes**, a description of **Use Codes for Parcels**, a field type of **Text**, and the type of domain as **Coded values**.

5   In the Code column, write **A1** with a description of **Single Family Detached**. Under that, write **A2** with a description of **Mobile Homes**. Continue down the form entering the rest of the Use Code values from the accompanying list. Print more worksheets if necessary.

A3	Condominiums	ESMT	Easement
A4	Townhouses	F1	Commercial
A5	Single Family Limited	F2	Industrial
B1	Multifamily	GOV	Government
B2	Duplex	POS	Public Open Space
B3	Triplex	PRK	Park
B4	Quadruplex	PROW	Private Right-of-way
CITY	Developed City Property	ROW	Right-of-way
CITYV	Vacant City Property	SCH	School
CITYW	Water Utility Property	UTIL	Utility
CRH	Church	VAC	Vacant

Domains worksheet

Domain name	Description	Field type	Domain type	Coded values / Range	
				Code (Min)	Desc (Max)
ParcelUseCodes	Use Codes for Parcels	Text	Coded values	A1	Single Family Detached
				A2	Mobile Homes
				A3	Condominiums
				A4	Townhouses
				A5	Single Family Limited
				B1	Multifamily
				B2	Duplex
				B3	Triplex
				B4	Quadruplex
				CITY	Developed City Property
				CITYV	Vacant City Property
				CITYW	Water Utility Property
				CRH	Church
				ESMT	Easement
				F1	Commercial
				F2	Industrial
				GOV	Government
				POS	Public Open Space
				PRK	Park
				PROW	Private Right-of-way
				ROW	Right-of-way
				SCH	School
				UTIL	Utility
				VAC	Vacant

Adding this domain will build in a check for data integrity. You can rest assured that the use code abbreviation entered for any piece of property will fit your normal list. Perhaps there are some other fields in the table that would benefit from the application of a domain. Most of them, such as a subdivision name or house number, couldn't be constrained in this way; there would be just too many values. But the field street type might be a good candidate. There is a standard set of street type abbreviations available from the U.S. Postal Service, and from time to time, you may be asked to generate a mailing list from this table. So, it would be a good idea to add a domain to this field.

There is a large number of acceptable street type abbreviations, and not only would you not want to have to list them all on the domains form, but you also wouldn't want to type them into a domain. Fortunately, there is a command to take a file listing of street types and read them into a domain. That process will be demonstrated later, but for now, the information about the file name can be recorded on the design worksheet.

6 On the tables worksheet, write **StTypeAbbrv** as the domain name for the field Street_Type and add a **(D)** noting that it is a domain.

7 Next, go to the domains worksheet and write the name **StTypeAbbrv**, a description of **Street Type Abbreviations**, a field type of **Text**, and a domain type of **Coded values**.

Domains worksheet							
Domain name	Description	Field type	Domain type	Coded values / Range Code (Min)		Desc (Max)	
StTypeAbbrv	Street Type Abbreviations	Text	Coded values	Data\Suffix.txt			

1-1

1-2

Under Code, write the file name **Data\Suffix.txt** to identify the file holding the domain values.

This file was created from a list on the U.S. Postal Service Web site, and it will be added into a domain later.

Now is a good time to investigate other aspects of how the data will be used and see if there are any other data integrity techniques that might be employed.

A component of the feature class will make built-in subcategories of the data. For instance, you might separate property as either platted (divided into developed lots with utilities) or unplatted (raw agricultural land). It is important to know the distinction for legal purposes and for the sale of property. It would be possible to put the platted land into one feature class and the unplatted property into another. If both of these feature classes were in the same geodatabase, they could both be edited at the same time. The symbology and annotation would work well for both, and each feature class could have different data integrity rules. So, dividing them into two feature classes would work and might be seen as beneficial.

But consider how the data might be used in a query. If a list of all property owners in a given region were needed, it would have to come from two separate files, and exporting the list would create two tables. So while it would be beneficial in some respects to put property data into two feature classes, using the data would be problematic. That's where subtypes come in.

A subtype is a way to subdivide data within the same feature class, and then apply different data integrity rules to each category. It's the best of both worlds: the data can be separated into logical categories and be given data integrity rules for each category but keep the convenience of being edited and managed in a single feature class. You'll also see later how subtypes can be used to set default values, establish unique attribute domains, set connectivity rules, and establish relationship rules for each subcategory created.

A field to carry the subtype code will need to be added to the table. The field type must be Integer, and the codes will be established along with a description. For this data, make a code 1 for Platted Property, a code 2 for Unplatted Property, and a code 3 for Plat Pending. This last code will be for property that has been approved by the city but is awaiting the filing data from the county. This will be a pretty simple subtype, without any additional data integrity rules added.

8 On the tables worksheet, write in a new field on the bottom called **PlatStatus**, make its field type **SI (short integer)**, give it an alias of **Plat Status**, and don't allow for null values. Since most new property being added to the dataset will be platted, make its default value 1. Finally, write the name **PlatSubtype** for the subtype name with a notation of **(S)**.

9 Next, go to the subtypes worksheet. Write the name of the subtype as **PlatSubtype** and add the three codes described previously:

- 1 = Platted Property
- 2 = Unplatted Property
- 3 = Plat Pending

Subtypes worksheet

Subtype name	Code	Description	PRESET DEFAULTS		
			Field	Domain name	Default value
PlatSubtype	1	Platted Property			
	2	Unplatted Property			
	3	Plat Pending			

Extend the data model

This concludes the initial design phase of the Parcels feature class, but there's another component to investigate. When these polygons are symbolized, they can each have a solid fill and a line style for their perimeter. When maps are made, however, the boundaries of the parcels will need to be symbolized differently. The edge of the parcel that fronts a street will be drawn with a thicker line; the edges representing property lines between properties will need to be a thinner line; and in the event that someone owns two adjacent pieces of property, the line between them should be dashed.

It isn't possible to symbolize polygons that way, so a solution needs to be developed. Consider creating a set of lines that will duplicate the boundaries of each parcel. Then these lines can be symbolized as described. The only field the feature class will need is a code describing which type of line to draw. This field would benefit from having a data integrity rule (a domain) with the three categories of lines described.

A behavior will need to be set up between the polygons representing property and the lines representing their boundaries. If the shape of any polygon is modified, the lines will need to automatically adjust to coincide. This type of relationship is called topology and will be discussed later (see chapter 5). In order for ArcGIS to manage this topology, the feature classes must reside inside the same feature dataset.

Feature datasets are another way to segregate data inside a geodatabase. If any behavior is to be built for a feature class, such as topologies, network databases, geometric networks, relationships, or terrains, they will have to reside in a feature dataset. For this example, you will need to establish a feature dataset for your feature classes, so that the corresponding topology can be built.

1-1

1-2

1 On the feature classes worksheet, write the name of the feature dataset as **PropertyData**. Next, write the new feature class name **LotBoundaries** on a blank line. Give it a feature type of **L** (for line) and an alias of **Lot Boundaries**.

Geodatabase name	LandRecords	
Feature dataset name	PropertyData	
Feature classes:		
Type Feature class name	Alias	
POLY Parcels	Property Ownership	
L LotBoundaries	Lot Boundaries	

2 Next, go to the tables worksheet and write the name of the new table as **LotBoundaries**. Then write the single attribute of this table, **LineCode**. Give it a field type of **Text**, add an alias of **Line Code**, and do not allow nulls. Note that there is a domain for this field and name it **ParcelLineCodes**.

3 Finish by filling in the information for the domain. On the domains worksheet, add the name of the domain **ParcelLineCodes**, a description of **Line Codes for Parcels**, a field type of **Text**, and note the domain type as **Coded values**. Then write the three domain values described previously:

- ROW = Right-of-way
- LOT = Lot Line
- SPLIT = Split Lot Line

Domains worksheet

Domain name	Description	Field type	Domain type	Coded values / Code (Min)	Range Desc (Max)
StTypeAbbrv	Street Type Abbreviations	Text	Coded values	Data\Suffix.txt	
ParcelLineCodes	Line Codes for Parcels	Text	Coded values	ROW	Right-of-way
				LOT	Lot Line
				SPLIT	Split Lot Line

Design a relationship class

The features you've dealt with in the design so far have been the points, lines, and polygons that will create the model of reality. Not all the data you will need for this model, however, is in the form of points, lines, and polygons. The design will also need to include tabular data that is provided by an outside agency. For each parcel, there is ownership and value information that comes from a county appraisal agency. This data would be valuable for analysis if it were associated with the parcel data. The nature of the table is that it is updated regularly from separate appraisal software, and because of this, it cannot be incorporated in the polygon feature class in the same way as regular data. By keeping it separate, it will facilitate the maintenance of both the ArcGIS use of the data and the third-party software's use of the data.

A relationship class has many of the benefits of a simple join in ArcMap but also provides a mechanism for controlling edits in the related table. If the graphic features were altered in an edit session, rules in the relationship class could also alter the related table and maintain the relationship. For this example, the parcels have a match in the appraisal roll table. If a piece of property is removed because of replatting, the associated record in the appraisal table can be set to be deleted automatically.

The final consideration is the cardinality of the relationship. If each parcel has one and only one match in the appraisal table, and vice versa, the cardinality is said to be one-to-one (1:1). If one parcel can have several matches in the appraisal table, such as the case of a single parcel being owned by more than one person, the cardinality is said to be one-to-many (1:M). If the opposite relationship were also true—that is, an owner can also own several pieces of property—the relationship is said to be many-to-many (M:M).

Relationships worksheet

Name of the relationship class _____

Origin table/feature class _____

Destination table/feature class _____

Relationship type   Simple (peer to peer)   Composite

Labels:
    Origin to destination _____
    Destination to origin _____

Message propagation   Forward Backward Both   None

Cardinality   1-1   1-M   M-N

Attributes   No   Yes - Table name _____
                                        Add to the tables worksheet

                   Primary key field   Foreign key name

Origin table/feature class _____ :

Destination table/feature class _____ :

Armed with this information, you can move to the worksheet on relationship classes and fill in the details.

1 On the relationships worksheet, name the relationship class **Ownership**. Record the origin table as **Parcels** and the destination table as **TaxRecords2010**.

Relationships worksheet

Name of the relationship class: **Ownership**

Origin table/feature class: **Parcels**

Destination table/feature class: **TaxRecords2010**

The relationship class can be used to add or delete records, but since the related table will be managed by another source, the relationship type should not allow records to be deleted, making it a simple peer-to-peer relationship. Labels will be shown to describe the relationship between the tables. The description for moving from the parcels feature class to the appraisal table is **Parcel is owned by** and from the appraisal table to the parcels feature class is **Owner has ownership of**. As the relationship is used in analysis, these labels will remind the user of the nature of the relationship. Normally, relationship classes are transparent (not visible to the user), but you can have ArcMap display a message when the relationship is used. For this example, opt not to use them.

2 Circle Simple (peer to peer) as the relationship type and write the labels **Parcel is owned by** for Origin to destination and **Owner has ownership of** for Destination to origin. Circle None for message propagation.

Next, note the cardinality as many-to-many, since a parcel can be owned by several people, and one person may own several parcels. It may also be beneficial to store what percentage of ownership can be attributed to each owner. This will help when more than one person is recorded as the owner. Write the name of the table as Ownership_Rel, and it will be added to the tables worksheet later. Finally, select the fields that will be the basis for the relationship and give them a label describing their relationship to the related table (foreign key description).

3 Circle M-N for cardinality on your design form, circle Yes under Attributes, and write the name of the relationship table as **Ownership_Rel**. Set the Origin and Destination primary key fields as Georeference. Name the Origin foreign key **Owner** and the Destination foreign key **Property**.

4 On the tables worksheet, add a new table called **Ownership_Rel** with a field called **PercentOwn** and a field type of **Float**. Write the alias as **Percentage Owned**, allow for nulls, and set the default value of **100**. This completes the logical model for the geodatabase. From these design forms, you will be able to create the entire structure using the ArcCatalog application and begin using it for storing data. If you do not have a lot of experience editing geodatabases, you may want to jump ahead to tutorial 2–1 and see how this design will function, and then come back to this exercise. Otherwise, complete this exercise, which will continue to focus on the design phase.

Exercise 1–1

The tutorial showed how to diagram a geodatabase to include feature classes along with their associated tables, domains, and subtypes. The goal was to think through the design, adding data integrity and behavior guidelines to the database.

In this exercise, you will repeat the process for another dataset required by the city planner. This one will contain the zoning data for Oleander. The zoning code for a piece of property determines the type of development that is allowed on a particular parcel. The zoning districts may incorporate several parcels and generally follow parcel boundaries.

The zoning districts will be represented by solid shaded polygons. The edges of the polygons will need to be symbolized in one of two ways—either as a solid line representing a zoning boundary or as a dashed line representing a change in allowable development density. Because of this, you will want to design a linear feature class for symbology purposes. The codes necessary for the zoning information are as follows:

R–1	Single Family Residential	C–2	Heavy Commercial
R–1A	Single Family Attached	TH	Townhouse
R–1L	Single Family Limited	LI	Limited Industrial
R–2	Duplex	I–1	Light Industrial
R–3	Triplex	I–2	Heavy Industrial
R–4	Quadruplex	TX–121	121 Development District
R–5	Multifamily	POS	Public Open Space
C–1	Light Commercial		

Analyze the descriptions of this data and determine what feature datasets and feature classes need to be made, what fields they should contain, any domains that might need to be created, and any subtypes that might be beneficial.

- Print a set of the geodatabase design forms as necessary.

- Use the forms to create the logical model for feature classes for the zoning polygons and the zoning boundaries.

- Investigate the use of domains and subtypes to build data integrity and behavior into your design.

WHAT TO TURN IN

If you are working in a classroom setting with an instructor, you may be required to submit the design forms you created in tutorial 1–1.

The completed geodatabase worksheets for

Tutorial 1–1
Exercise 1–1

Tutorial 1–1 review

Over the last thirty years, the manner in which geographic features have been portrayed, stored, and manipulated in GIS has evolved from a file-based technology into the present-day ESRI geodatabase format. By using the ESRI geodatabase, GIS practitioners can more realistically manage geographic features and their relationships to other features. Although computer technology has enhanced the behavioral aspects of these relationships, the fundamental ways that these geographic features are represented—by points, lines, and polygons—has largely remained unchanged. ESRI geodatabase technology has improved the management of these points, lines, and polygons by providing tools to create geographic feature representations, enforce data integrity, and establish relationships among the geographic features that more closely model real-world situations.

As illustrated in the previous exercise, the opportunities to manage data using GIS methodology can be enhanced by careful thought and preplanning to ensure that an accurate portrayal of geographic features and their relationships is contained in the geodatabase. Preplanning the geodatabase is enhanced through the use of a structured, organized logical data model to ensure that every conceivable relationship is accounted for in the model. This preplanning phase is no easy task. However, it is much easier to spend time at the outset designing your geodatabase than it is to change it once you've begun entering data into the model.

Organizing your geodatabase, through the use of feature classes and feature datasets, allows you to refine relationships and behaviors for the data. **Feature classes**, as the most basic representation of geographic data in the geodatabase, can be logically grouped together to form **feature datasets**. Although there are many different techniques for organizing geographic data in the **geodatabase**, the organization of the data must be guided by the behavior of these features in the real world. For example, if feature classes contained in the geodatabase work together to form a **geometric network**, represent a **terrain**, or establish a **topology**, the feature classes must reside in the same feature dataset. Such behaviors among the data must be considered while designing the geodatabase.

Once your design is complete, using domains for your attribute data and other techniques will reduce costly mistakes during the data entry phase of development of your geodatabase. Additional techniques provided by the geodatabase such as the creation of **subtypes** optimize how data is organized and used within the geodatabase. Through the use of the ESRI geodatabase, the many tools available within ArcMap and ArcCatalog, and a thoroughly planned approach, your new geodatabase will adequately portray the geographic features and associated relationships among them. As a result, your model of reality as contained in the geodatabase will represent the real-world features as closely as possible.

STUDY QUESTIONS

1. What is a logical model of a geodatabase, and why should you develop a logical data model when designing your geodatabase?
2. What are the principal advantages of using subtypes? Give one example of a situation where you would create a subtype and specify why.
3. What are the principal advantages of using domains?
4. What is the difference between feature classes and feature datasets? When must you create a feature dataset?

Other real-world examples

Techniques described in this tutorial to create and manage geographic information in a geodatabase can be applied to a wide variety of situations to successfully simulate real-world circumstances. Any situation that involves finding and mapping the location of a person, building, or event will benefit from a GIS approach.

Water utilities around the world use GIS methodology to help manage their infrastructure. Water pipeline systems contain numerous components that often include pipe segments, valves, and joints, as well as information on certain soil conditions around each pipe that affect pipe integrity. As a GIS analyst, you may be tasked with the development of a geodatabase to represent each of these features, as well as to model the critical behaviors and relationships of these components. All the features can be represented by points, lines, and polygons, but to be successful, there will need to be a significant effort in duplicating the real-world relationships and behaviors in the geodatabase.

Electrical utilities also use GIS methodology to help manage vast networks of electrical grids. A principal use of GIS by a power company may involve developing inventories of all the assets in its power grid. The inventory may consist of multiple types of power lines, power poles, and transformers. All these assets also have specific behaviors when relating to other assets as well as to customers who rely on the network to provide reliable and cost-effective electricity. A properly designed geodatabase allows for the assessment of critical infrastructure features and provides an opportunity to accurately model the relationships and behaviors among the features in the geodatabase.

Once you become familiar with the advantages of the data creation and management tools available in the geodatabase, you will be able to simulate virtually any geographic situation and use these features to successfully manage the resulting geodatabase to your advantage whether it be keeping track of utility assets, satisfying customer demands, or any number of other GIS-related tasks.

Tutorial 1-2

Creating a geodatabase — expanding the logical model

The components of a geodatabase can have various spatial relationships, or behaviors, that form a topology. These behaviors can exist among points, lines, and polygons and will have an impact on the logical model for a database. The most efficient designs will consider topology from the beginning.

Learning objectives

- *Design linear feature classes*
- *Investigate data behavior*
- *Design for topology*
- *Design point feature classes*

Introduction

The first tutorial used the geodatabase design forms to construct a logical design for a parcels database. That dataset consisted of a polygon feature class along with a linear feature class to aid in symbolizing the parcel boundaries.

In this tutorial, you will design another set of feature classes to store data for a sewer system. The process will include investigating the behavior of the data, and then trying to accommodate it in the design.

Remember to look at how the data will interact with feature classes, as well as any possible domains or subtypes that may be used. This will help to build not only an efficient design but also a good model of reality.

Designing the data structure

Scenario   After your successes with the parcels and zoning datasets, the Oleander Public Works Department is seeking your help to create a geodatabase for the sewer system. You will need to design this for them.

1-1

1-2

Sewer systems are rather simple. They consist of pipes to carry wastewater to the treatment plant. In real life, it's important that the pipes connect to ensure a direct flow route from the beginning of the system to the end. In the data model, you will also want to ensure connectivity, which will allow the data to be used later to construct a network dataset.

A great amount of data, such as the size of the pipe, the material it is made of, and the year it was installed, can be stored as attributes of the linear features.

This portion of the process is intended to inspire thought and creativity. If done correctly, your designs will be valuable for years to come. Once all the designs are completed, they will be used to create the data structure in ArcCatalog.

Data   Since you are creating this geodatabase from scratch, there is no data to start with. But you will need to print the geodatabase design forms on the DVD as an aid in the design process.

Tools used   Geodatabase design forms

Begin the geodatabase design process

Using the geodatabase design forms, you will once again commit your thoughts to paper and examine all aspects of how the data will be used, edited, and symbolized. The first part of the design will be to name the geodatabase. Since the data, in all likelihood, will be used in a network later, it will also require a feature dataset.

1 On the first page of a new set of design forms, write the name of the new geodatabase as **Utility Data**. On the next line, add the name of the feature dataset as **Wastewater**. The sewer lines will be built as linear features, which will require a feature class. A number of attributes could also be stored with the lines, as mentioned earlier. Add each of these to the design forms.

2 On the geodatabase design form, write the name of the new feature class. Give it the name **SewerLines** with a feature type of **L** and an alias of **Sewer Lines**.

Next, you will need to fill in the tables worksheet and show which fields the feature class will contain. The three fields that were required by Public Works were pipe size, which can be a number; pipe material, which can be text; and the year the pipe was installed, which is also a number.

3 On the tables worksheet, write the name of the feature class. Then write in the fields **PipeSize** with a data type of **SI**, **Material** with a data type of **Text**, and **YearBuilt** with a data type of **LI**. Add the aliases of **Pipe Size**, **Pipe Material**, and **Year Built**, respectively. In Oleander, there are sewer lines that run through the city that belong to other agencies. Some are the pipes of other cities and are headed for the treatment plant, and some belong to the regional utility that handles all the wastewater treatment for local cities. They will all need to be included in the dataset, and on the maps, to prevent accidentally digging into them. The owner of the line needs to be recorded, so you'll add a field called Description to store the name of the owner of each pipe.

4 In the field name column, add a field called **Description**. Write a field type of **Text** with an alias of **Owner**.

Tables worksheet						
Feature class or table name	Field name	Field type	Alias		Nulls (Y/N)	Default value	Domain name or subtype field (D) or (S)
SewerLines	PipeSize	SI	Pipe Size			
	Material	Text	Pipe Material			
	YearBuilt	LI	Year Built			
	Description	Text	Owner			

Data integrity issues

With this data, it is important that every pipe have an entry for size and material. Year of construction, however, may not be known for some of the older, existing pipes. Because of this, do not accept null values for the fields PipeSize and Material, but allow nulls for YearBuilt. Also, the ownership of every pipe must be known, so don't allow for nulls.

1 For each field, write **No** next to the aliases Pipe Size, Pipe Material, and Description in the Nulls column. Write **Yes** next to Year Built in the same column. The next data integrity issue will be to investigate the use of domains. Sewer pipes vary in size from 6 inches to 12 inches in 2-inch increments. Pipes larger than 12 inches are called interceptors and are metered to determine the charge to the city. In Oleander, the interceptors are owned by a regional utility that handles all the wastewater processing. While they run through the city, Oleander's Public Works Department does no service or maintenance on them.

If a domain were built for pipe size, it could prevent some data entry errors. The choices would be to use coded values and enter a discrete set of values or use a range and give a low and high value such as 6 and 12. A range would allow any numeric entry between these values, and since the sizes increase in 2-inch increments, there would be values allowed by the domain that are not allowed in reality. For example, the range from 6 to 12 would allow an entry of 9, but there is no such thing as a 9-inch sewer pipe. So, this wouldn't work. It is apparent that a discrete list of coded values will need to be entered.

2 On the tables worksheet, write the name of the domain for the field PipeSize as **SewerPipeSize** with the **(D)** notation. Then on the domains worksheet, write the same name. Add a description of **Sewer Pipe Size**, set the field type as **SI**, and write the domain type as **Coded values**. Enter the values as shown and their corresponding descriptions:

- 6 = 6"
- 8 = 8"
- 10 = 10"
- 12 = 12"

Domains worksheet

Domain name	Description	Field type	Domain type	Coded values / Range Code (Min)	Desc (Max)
SewerPipeSize	Sewer Pipe Size	SI	Coded values	6	6"
				8	8"
				10	10"
				12	12"

Notice that although the field stores integers, and the code must be an integer, the associated description can be text. This will be useful in labeling the text later, as the inch marks will be visible on the labels that ArcMap generates.

Another data integrity tool is to include subtypes. These can be used to segregate data, so that there will be separate domains and defaults for each subset of data. In this scenario, the data might be separated by material. Almost all the new polyvinyl chloride (P.V.C.) pipes going in are 8 inches, almost all the new high-density polyethylene pipes are 10 inches, and almost all the ductile iron pipes going in are 12 inches. These are the only materials allowed for new pipes, so if each of these were set up as a subtype, additional control could be added to automatically populate some of the more common fields.

One problem with this approach would be the interceptors. These are typically larger than 12 inches, but the size and material change for each situation. Default values wouldn't be appropriate here, so the interceptors don't play by the same rules as the Oleander pipes. Perhaps the solution is to put them in their own feature class. They could still be edited simultaneously with the Oleander data, and still participate in any networks that are built, provided that they reside in the same feature dataset. Also, the fields for interceptors would be identical to the Oleander lines. Update the worksheets to include an additional feature class for interceptors.

3 Add the name of the new linear feature class as **Interceptors** on the geodatabase worksheet and give it an alias of **Interceptors**. Be sure to fill in its type as **L**.

4 Write the name of the feature class on the tables worksheet and duplicate all the fields and properties from the SewerLines feature class. Since the interceptors don't have any regular size or material, there will be no domains or default values for these lines. With the problem solved, you can proceed in designing the subtypes. A good field using a subtype is Material. By selecting the material, the default values will automatically populate the other fields. And in the event that a size other than the standard is used, the pipe size domain will prevent any incorrect values from being entered.

The subtype field must always be an integer, and the material field is set as text. This can be changed easily with an eraser.

5 On the tables worksheet, erase the field type for Material and enter **SI**. Also, add the name **SewerLineMaterial** in the subtype column at the right with an **(S)** notation.

Tables worksheet

Feature class or table name	Field name	Field type	Alias	Nulls (Y/N)	Default value	Domain name or subtype field (D) or (S)
SewerLines	PipeSize	SI	Pipe Size	No		(D) SewerPipeSize
	Material	SI	Pipe Material	No		(S) SewerLineMaterial
	YearBuilt	LI	Year Built	Yes		
	Description	Text	Owner	No		

Next, you can fill in the subtypes worksheet for the first material, polyvinyl chloride (commonly abbreviated P.V.C.). The default value for PipeSize will be 8 inches, and the domain designed previously needs to be applied to this field. The default for the description field will be Oleander. And to save a little typing, make the default for YearBuilt 2010. You can change it once a year to keep up with new construction.

6 On the subtypes worksheet, write the subtype name **SewerLineMaterial**. Write the code as **1** and the description as **P.V.C.** In the field column, write the name **PipeSize**, note its domain as **SewerPipeSize**, and add its default value of **8**. Note that the default value does not have the inch marks next to it. Remember that its field type is short integer, so the value entered in the database must be a short integer. But the inch mark was stored in the domain description, which can be used for labeling if necessary.

7 Now, write the names of the other fields from the SewerLines table and their default values. For the field named Description, write a default value of **Oleander**. For the YearBuilt field, write a default value of **2010**. This completes the design for the first choice of subtype. The next choice will be for the material type of high-density polyethylene, or HDPE. The default size will be 10 inches, and the defaults for description and year built will be the same as before.

8 On the next blank line of the subtypes worksheet, write the code of **2** with a description of **HDPE**. In the field column, write the name **PipeSize**, note its domain as **SewerPipeSize**, and its default value as **10**. As before, write a default value of **Oleander** for the Description field, and **2010** as the default value for the YearBuilt field.

Subtypes worksheet				PRESET DEFAULTS	
Subtype name	Code	Description	Field	Domain name	Default value
SewerLineMaterial	1	P.V.C.	PipeSize	SewerPipeSize	8
			Description		Oleander
			YearBuilt		2010
	2	HDPE	PipeSize	SewerPipeSize	10
			Description		Oleander
			YearBuilt		2010

YOUR TURN

Fill in the information for the material type ductile iron, or DI. It will have a default size of 12 inches with the same domain as the other sizes, as well as a default description of Oleander and default year built of 2010.

There are two more material types, and although they are no longer installed new, they could cause validation problems later if they are not included in the design. The two types are concrete and clay. Add them as choices 4 and 5 on the subtypes worksheet. They will require no domains or defaults since they will not be used to enter new pipes.

9 On the next blank line, write a code **4** with a description of **Conc** and a code **5** with a description of **Clay**. No defaults or domains are required for these. This completes the design for linear features. Next will be to investigate the points features associated with the sewer lines. At each intersection of sewer lines, and at various locations along their length, manholes are constructed for maintenance. At the ends of the lines, a smaller access port called a cleanout is added to accommodate the mechanical device that is run down the pipes to clean out clogs. The cleanouts will be represented in the geodatabase by points, with certain attributes associated with them.

These points have a behavior relationship with the lines in that they must fall on top of the lines. If any networking is done, the points will need to be snapped to the lines to preserve connectivity. They will also need to reside in the same feature dataset.

The data associated with the points will include several fields. The first will be a code marking which points are manholes and which are cleanouts. Other information such as the flow line, rim elevation, and depth (rim elevation minus the flow line elevation) will be copied from the construction documents. Finally, a field for the year of construction and another for description (ownership) will be needed.

10 On the geodatabase design form, write the name of the new point feature class as **SewerFixtures**. Give it a feature class type of **PNT** (for point) and an alias of **Sewer Fixtures**.

Next, add the fields for the point feature class on the tables worksheet.

11 On the tables worksheet, write the name of the new layer and add the following names with their field types, aliases, and null value allowances:

- FixType, SI, Fixture Type, No
- Flowline, Float, Flow Line Elevation, Yes
- RimElev, Float, Rim Elevation, Yes
- Depth, Float, Depth from Surface, Yes
- YearBuilt, LI, Year Built, Yes
- Description, Text, Owner, No

Even though there won't be defaults or domains for any of these fields, it might be useful to make fixture type a subtype. One benefit of subtypes is that each code in the subtypes list would be added as a separate selection on the editing template when you start adding new features. Without a subtype, you would get a single template entry for SewerFixtures and you would use it to add a point. As points were added, you would have to set the fixture type immediately since it cannot be null. With a subtype set for fixture, the editing template would show the two types of fixtures allowed. To add new fixtures, you would select which fixture you wanted from the templates and the fixture type field would be automatically populated, meaning that it can never be null. For the short amount of time it takes to set up the subtype structure, this would be a great way to enforce the data integrity rule of not allowing null values. At the same time, a default for description and year built could be added for convenience. Start by noting the subtype name on the tables form and then populate the subtype form..

12 On the tables worksheet, write the subtype name **SewerFixType** for the FixType field with a notation of **(S)**. Add a default YearBuilt value of **2010** and a default Description value of **Oleander**.

Tables worksheet

Feature class or table name	Field name	Field type	Alias	Nulls (Y/N)	Default value	Domain name or subtype field (D) or (S)
SewerFixtures	FixType	SI	Fixture Type	No		(S) SewerFixType
	FlowLine	Float	Flow Line Elevation	Yes		
	RimElev	Float	Rim Elevation	Yes		
	Depth	Float	Depth from Surface	Yes		
	YearBuilt	LI	Year Built	Yes	2010	
	Description	Text	Owner	No	Oleander	

13 On the subtypes worksheet, write the name of the subtype as **SewerFixType**. Give it a code of **1** for Manhole and a code of **2** for Cleanout. The interceptors will also have associated fixtures, but they are all manholes. This makes for a very simple feature class since all the features will be symbolized the same. They will have the same fields as the Oleander fixtures, except for fixture type and description. These fields are unnecessary since their values would always be the same.

14 Add a new feature class to the geodatabase worksheet. Name it **InterceptorFix**, with a feature type of **PNT** and an alias of **Interceptor Fixtures**.

15 Finally, fill in the tables worksheet for the new feature class InterceptorFix with these field names, field types, aliases, and null values:

- FlowLine, Float, Flow Line Elevation, Yes
- RimElev, Float, Rim Elevation, Yes
- Depth, Float, Depth from Surface, Yes
- YearBuilt, LI, Year Built, Yes (default value of 2010)

The geodatabase design forms make this design seem pretty simple, but in reality, it is a fairly complex database. A good deal of thought was put into the fields required for the feature classes, the relationships of the feature classes, and the inclusion of data integrity rules such as defaults, domains, and subtypes.

Review the design forms and resolve any questions that you may have, because the next tutorial will have you build these data structures in ArcCatalog.

Exercise 1–2

The tutorial showed how to apply design strategies and data integrity rules to point and linear feature classes. Each feature type was analyzed against the reality it is supposed to model to try and build in as much behavior and data integrity as possible.

In this exercise, you will repeat the process with storm drain data. The City of Oleander's Public Works Department would like a geodatabase design for the storm drain system just like the one you did for the sewer collection system. It will consist of the pipelines and fixtures associated with them. The lines are all made from reinforced concrete pipe (RCCP) and vary in size from 15 inches to 45 inches in 3-inch increments. The pipes are usually classified as laterals (21 inches or less), mains (more than 21 inches), and boxes (square pipe with no restriction in size). The data for these features includes a pipe size, material, description, flowline in, flowline out, slope, year installed, and a designation for public or private line.

Connected to these pipes are various types of fixtures listed as follows. These will be used as the subtypes, and their code from the existing data is included:

101 = curb inlet

102 = grate inlet

104 = junction box

106 = Y inlet

107 = junction box/manhole

108 = outfall

109 = headwall

110 = beehive inlet

111 = manhole

The type of data collected for these features includes a description, flowline elevation, inlet size, top elevation, year built, designation for public or private line, and a rotation angle.

Analyze the descriptions of this data and determine what feature datasets and feature classes need to be made, what fields they should contain, any domains that might need to be created, and any subtypes that might be beneficial.

- Print a set of the geodatabase design forms as necessary.
- Use the forms to create the logical model of feature classes for the zoning polygons and zoning boundaries.
- Investigate the use of domains and subtypes to build data integrity and behavior into your design.

Getting started

Here's a little help to get you started:

- Decide how many feature classes you want to make.

- List the fields that will need to be in each feature class.

- Determine the field type, null status, and default value for each field.

- Investigate the use of domains for these fields.

- Look for fields that describe a "type" or "category" that could be used as a subtype, such as fixture type.

1–1

1–2

WHAT TO TURN IN

If you are working in a classroom setting with an instructor, you may be required to submit the design forms you created in tutorial 1–2.

The completed design worksheets for

       Tutorial 1–2

       Exercise 1–2

Tutorial 1–2 review

While the first tutorial focused on the development of a geodatabase to contain parcel-related information, this tutorial focused on the development of a geodatabase to represent a sewer system. Both tutorials focus on real-world examples of the types of critical information managed by every city, town, or other local governmental entity. Given the tremendous municipal resources dedicated to the management of these systems, an adequate foundation for managing them is essential. The geodatabase allows for physical information, as well as information on the behaviors among the components of these systems, to be accurately portrayed in a GIS. This, in turn, enhances effective and efficient decision-making capabilities.

One of the most critical steps in developing a comprehensive geodatabase is the initial planning phase. While we have focused on and discussed GIS functionality within both tutorials, we have yet to even open ArcGIS! Planning for a geodatabase development project involves both gathering the physical system requirements and understanding how interrelated objects behave in the physical system. To further enhance your design, you will also need to know what kinds of questions your customers will need to have answered. Once an adequate knowledge of the system is gained, it is then up to the skill of the geodatabase developer to build these capabilities into the corresponding geodatabase.

A properly designed geodatabase will be worthless to the engineers and others who rely on it to represent the real world if it is full of errors. The geodatabase allows the designer to apply **domains** to feature attributes to ensure that the correct information is correctly recorded within the geodatabase. Flexibility to adequately portray and control features and their behavioral characteristics within the model is afforded through the use of **subtypes** of features. Interrelated behavior among features is enhanced through the use of **topology**. The success of a geodatabase often depends on a thorough knowledge of how and when to apply data integrity tools.

STUDY QUESTIONS

1. What is topology, and why is it an important concept when designing a logical model of a geodatabase?

2. Think about the principal ways in which features are represented spatially within a GIS. Give an example of each feature type, a possible domain for each type, and possible subtypes for each. Explain your rationale.

3. Why is it important to fully understand a system to be represented and managed in a GIS? How do you determine its structure?

Other real-world examples

Linear networks are commonly represented within a geodatabase. Applications range from water and sewer systems such as those illustrated in this tutorial to stream networks that help manage our natural resources to infrastructure that manages the national highway system.

Environmental agencies across the country use a national geodatabase to enhance the management of water resources in urban and natural environments. By using topological relationships among stream reaches, these agencies use the geodatabase to help alleviate flooding while maintaining the natural aspects of these systems to ensure environmental integrity.

Transportation agencies use the power of GIS to manage thousands of miles of streets and highways across the country. Collectively, these agencies maintain a variety of information related to road usage and conditions, traffic conditions, associated roadside facilities, and road signage. Relationships among these interrelated geographic features can be managed efficiently within GIS by establishing topology, using domains, and creating subtypes within a geodatabase.

References for further study

There are a number of additional resources to help design a framework for a complex geodatabase. **ArcGIS Desktop Help** lets you use keywords to search for a title or topic. You can access ArcGIS Desktop Help in ArcMap or ArcCatalog by clicking Help on the main menu. Then click the Search tab, type a keyword, and click Ask. Use the keywords provided here to search for additional learning resources on many of the concepts taught in this chapter.

ESRI has developed a number of online courses on a wide variety of pertinent GIS topics. **ESRI Self-Study (Virtual Campus) courses** are an excellent resource for students that will supplement information covered in this course. ESRI Virtual Campus courses can be accessed at `http://training.esri.com/gateway/index.cfm`.

ArcGIS Desktop Help search keywords

geodatabase, attribute domains, feature class, feature datasets, working with geodatabases, feature class basics, getting started with ArcGIS Desktop, subtypes, data integrity, topology

ESRI Self-Study (Virtual Campus) courses

The following ESRI Virtual Campus classes may be helpful:

1. Basics of the Geodatabase Data Model
2. Creating, Editing, and Managing Geodatabases for ArcGIS Desktop
3. Working with Geodatabase Subtypes and Domains
4. Introduction to ArcGIS Data Models

2

Creating a geodatabase

Tutorial 2–1

Building a geodatabase

Designing a geodatabase can be a long and drawn-out process, but it's only after that phase is completed that the data structure can be created in ArcCatalog. A good design will greatly simplify the creation phase and make it go more smoothly and quickly.

Learning objectives
- *Work with ArcCatalog*
- *Create a geodatabase*
- *Build a database schema*

Introduction

A good database design is essential to smooth creation of a geodatabase. It is best to think through the entire design, documenting your needs and addressing them with the logical model. Once that is completed, the creation phase can begin.

Databases can be very sensitive to change, and in fact, some elements, once created, cannot be changed. They can only be deleted, and then re-created. It is therefore very important when going through the creation tools to pay attention to what has been designed. Creating a feature class without regard to the data type will result in wasted time since the data type cannot be altered later. The same will hold true for field names and null allowances. Once created, these things cannot change.

The order of creating components is not necessarily as critical. Feature classes created outside a feature dataset can usually be moved into a feature dataset with a simple drag-and-drop operation. Field names left out of tables can usually be added later, or subtypes can be set up after the fact.

But even though some alterations can be made to the data structure later, it is best to have a complete logical model prepared and to follow it closely when creating the geodatabase. At this point, all opinions should have been heard and all aspects of the design completed. Major alterations to a data schema after the fact may require exporting the data to new feature classes, which could end in errors and confusion.

Creating the data structure

Scenario   The City of Oleander has passed along completed geodatabase design forms and asked that you build the data structure for them. The forms show feature datasets, feature classes, fields, domains, and subtypes. They have some existing data that will be loaded into this schema later.

Data   Print out the design forms provided. These are the completed geodatabase logical model diagrams from the previous tutorial.

2–1

2–2

Tools used   ArcCatalog:

       New Personal Geodatabase

       New Feature Dataset

       New Feature Class

       Create a Domain

       Table to Domain

As you go through the creation process, pay close attention to all the steps involved. Sometimes setting even one option wrong can result in having to delete the entire piece and start over. The order of creation is not as critical. Some people may prefer to make all the domains first, then create all the tables, and follow that up with subtype creation. Others will create one feature class, its table, domains, and subtypes, and then move on to the next table. As long as the components are understood and created correctly, the order is not important.

Create the data structure

You are starting with a blank canvas and will be creating everything from scratch. The first step will be to create the geodatabase, which will contain all the rest of the components. This could be an SDE geodatabase or file geodatabase depending on your particular setup, but for ease and clarity, this tutorial will create a personal geodatabase.

1   Print the geodatabase design forms from tutorial 1–1. If you did not complete the tutorial, or if you are not confident in your answers, a completed set of forms can be found in the Data folder called Tutorial 1–1 GDB Design.

Geodatabase name		**LandRecords**	
Feature dataset name		**PropertyData**	
Feature classes:			
Type	Feature class name	Alias	
POLY	**Parcels**	**Property Ownership**	
L	**LotBoundaries**	**Lot Boundaries**	

2 Start ArcCatalog and connect to the folder where you installed the tutorial data (for example, C:\ESRIPress\GIST3\). Right-click the MyAnswers folder, select New, and click File Geodatabase.

3 Type the name of the database that's on the design form, **LandRecords**.

Create a feature dataset

From the design form, you can see that there will be a feature dataset in this geodatabase, with two feature classes. The geodatabase itself only needed a name to exist. The feature dataset needs two things: a name and a spatial reference.

The spatial reference is what ties your data to a location on the globe. You may need to decide if your data needs to be stored as projected or unprojected data, and you will need to define the spatial extent of your dataset. This dataset needs to cover the City of Oleander in North Central Texas and will use a spatial reference that is typical for the area. For more information on selecting the spatial reference for other datasets, see the section in ArcGIS Desktop Help titled "An overview of spatial references."

1 Right-click the geodatabase, select New, and click Feature Dataset.

2 Type the name of the feature dataset as **PropertyData** and click Next.

The next screen is for setting the spatial reference. Oleander is in North Central Texas, and you will scroll through the lists for projected data to find the correct spatial reference. The one used for Oleander is Texas State Plane, NAD 1983 (feet), for North Central Texas.

3 Double-click Projected Coordinate Systems, then State Plane, then NAD 1983 (Feet), and then scroll down and find Texas North Central. Click Next and accept the default values for the next two screens, clicking Finish to complete the feature dataset.

Create the feature classes

Inside this feature dataset, you will create two feature classes. A feature class, at the minimum, needs three things: a name, a spatial reference, and a geometry type. You will provide the name from the design form. The spatial reference has been set at the feature dataset level, and the feature class will inherit this spatial reference. The final thing will be to set the geometry type. Pay close attention to set this correctly as it cannot be changed later.

2-1

2-2

1   Right-click the PropertyData feature dataset. Select New, and click Feature Class.

2   Type the name and alias. Then click the data type drop-down arrow, and click Polygon Features. Click Next, and then click Next again to accept the default option for Configuration Keyword. The groundwork for the feature class has been created. Next is to enter all the fields where the attributes will be stored. Refer to the tables worksheet for the names and configurations of the fields. As you enter each field, check the worksheet to get all the settings correct. Then highlight the row and double-check all the entries before moving on to the next field. This will help you avoid mistakes, which could cause you to have to delete the feature class and start over.

Tables worksheet

Feature class or table name	Field name	Field type	Alias	Nulls (Y/N)	Default value	Domain name or subtype field (D) or (S)
Parcels	SubName	Text	Subdivision Name	No		
	Blk	Text	Block Designation	No		
	LotNo	Text	Lot Number	No		
	Pre_Type	Text	Prefix Type	Yes		
	Pre_Dir	Text	Prefix Direction	Yes		
	House_Num	Text	House Number	No		
	Street_Name	Text	Street Name	No		
	Street_Type	Text	Street Type	No		(D) StTypeAbbrv
	Suffix_Dir	Text	Suffix Direction	Yes		
	ZIPCODE	LI	ZIP Code	No		
	UseCode	Text	Land Use Code	No		(D) ParcelUseCodes
	Georeference	Text	Georeference	No		
	PlatStatus	SI	Plat Status	No	1	(S) PlatSubtype

3 On the first blank line, type the field name **SubName** and set the data type to Text. In the Field Properties pane, type the alias **Subdivision Name** and use the drop-down arrow to set "Allow NULL Values" to No. Be careful and methodical as you work your way through the field entry process. It is desirable to do this in one pass through the dialog box to avoid getting confused or forgetting to set a value. Once the Finish button is pressed, only the alias can be later changed. Mistakes must be deleted and the files constructed again.

4 On the next blank line, type the field name **BLK** and set the field type to Text. In the Field Properties pane, type the alias Block Designation, and use the drop-down arrow to set "Allow NULL values" to No.

YOUR TURN

Work your way down the list of fields from the tables worksheet and enter all the values for field name, data type, alias, and Allow NULL values. At this point, disregard the default value and domain/subtype entries. If you run out of blank lines for new fields, use the slider bar at the left to reveal more. When all the fields have been entered, click Finish.

The feature class has been created and added to the feature dataset. There's one more feature class to create for the lot boundaries. The process will be the same, but this feature class only has one attribute field, so it will go much faster.

5 Right-click the PropertyData feature dataset. Select New and click Feature Class.

6 Type the name and alias as shown on the design form. Then click the data type drop-down arrow and select Line Features. Click Next, and then click Next again to accept the default setting for Configuration Keyword. Refer to the tables worksheet for the names and configurations of the fields.

Tables worksheet						
Feature class or table name	Field name	Field type	Alias	Nulls (Y/N)	Default value	Domain name or subtype field (D) or (S)
LotBoundaries	LineCode	Text	Line Code	No		(D) ParcelLineCodes

7 On the first blank line, type the field name **LineCode** and set the field type to Text. Add the alias of **Line Code**, and use the drop-down arrow to set "Allow NULL values" to No. Click Finish.

Create the domains

Now you have both feature classes, with all their fields. The next step will be to introduce data integrity rules for the feature classes. This will start with creating the domains and then assigning them to the fields in the feature classes. Domains are stored at the geodatabase level, which means that all the required domains for all the feature classes residing in the geodatabase will be stored there. You may be working on a utility project, but there may be others with their own feature datasets and feature classes requiring dozens of domains. Because of this, there are two critical rules when working with domains.

The first rule is to give your domain a very specific name, even if it means making it rather long. A domain called "Type" may be confusing, because it doesn't describe its purpose very well. Is this a type of road, sign, tree, or something else? A better name might be "TypeOfSignpost," if it were used to constrain the entry of signposts to certain types such as metal or wood.

The second rule is to never alter someone else's domain to fit your needs. You may find a domain called "Material" with a list of wood, steel, fiberglass, and concrete. The domain you want may be the exact same list but without concrete, so you delete that entry and use it. The next time the dataset for which that domain was originally created is used, all the features with a material of concrete will be invalid with respect to the material domain because the entry is missing. If it gets changed back to make that data valid, the new dataset will become invalid when referencing that domain. To prevent this from happening, it's best to always create your own domains.

The domains worksheet will be the guide for creating domains, so get that ready.

Domains worksheet

Domain name	Description	Field type	Domain type	Coded values / Range	
				Code (Min)	Desc (Max)
ParcelUseCodes	Use Codes for Parcels	Text	Coded values	A1	Single Family Detached
				A2	Mobile Homes
				A3	Condominiums
				A4	Townhouses
				A5	Single Family Limited
				B1	Multifamily
				B2	Duplex
				B3	Triplex
				B4	Quadruplex
				CITY	Developed City Property
				CITYV	Vacant City Property
				CITYW	Water Utility Property
				CRH	Church
				ESMT	Easement
				F1	Commercial
				F2	Industrial
				GOV	Government
				POS	Public Open Space
				PRK	Park
				PROW	Private Right-of-way
				ROW	Right-of-way
				SCH	School
				UTIL	Utility
				VAC	Vacant

1 Right-click the geodatabase LandRecords, and then scroll down and click Properties.

2 Click the Domains tab. This is where domains are created. On the first empty line, type the domain name **ParcelUseCodes** and add the description **Use Codes for Parcels**.

3 In the Domain Properties pane, set the field type to Text. Remember that there are two types of domains. The first is coded values, which contains a list of values. When you set the field type to Text, the only choice is Coded Values, so it is set automatically. Numeric fields can also be used with the Coded Values choice. The other type of domain is a range. This lets you set a start and end value within which the entered data must fall, but it is only valid with numeric field types.

4 Next, move down to the Coded Values pane. On the first line in the Code column, enter the first value of **A1**. Next to it in the Description column, enter the value **Single Family Detached**. Click Apply to save the entry.

YOUR TURN

Using the domains worksheet, enter the rest of the values for code and description. Periodically, click Apply to save your work and make sure there are no errors in the domain. When you have completed the list, click OK.

There are two more domains to make. One will use the same process as the use codes domain, and the other will use a process to import the values from an existing database. Create the parcel line codes domain first, since it uses the same dialog as the previous example.

5 Right-click the geodatabase LandRecords, select Properties, and click the Domains tab. You'll see the domain created earlier, as well as any other domains that anyone else may have made in this geodatabase. Click the first empty line and type the new domain name **ParcelLineCodes** as well as the description **Line Codes for Parcels**.

6 Move to the Domain Properties pane and set the field type to Text. The domain type will automatically set to coded values. In the Coded Values pane, enter the codes and descriptions for the domain:

- ROW = Right-of-way

- LOT = Lot Line

- SPLIT = Split Lot Line

When all the values are entered, click Apply, and then click OK.

Create a domain from a table

The next method of domain creation will use an existing database of street suffixes to fill in all the values. In the ArcToolbox interface, there is a tool that will read the table and transfer it into the domain you created in the geodatabase. The first step will be to find the tool.

1 In ArcCatalog, use the Search tool to locate the Table to Domain tool.

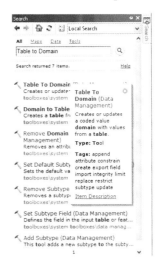

2 Double-click Table to Domain to start the tool. In the dialog box, click the Browse button next to Input Table. Browse through the C:\ESRIPress\GIST3\Data folder for the table Suffix.txt. Click the table, and then click Add.

3 In the Code Field input box, click the drop-down arrow and select SuffixAbbrv.

4 In the Description Field input box, select SuffixType.

5 Use the Browse button to set Input Workspace to the LandRecords geodatabase, and then click Add.

2–1

2–2

6 Finally, type the domain name as **StTypeAbbrv** and description as **Street Type Abbreviations**. When your dialog box matches the accompanying image, click OK. The domain was created with all the suffix values that are used by the U.S. Postal Service. This is a quick and easy option for creating domains that come from other data sources.

7 Open Properties of the geodatabase in ArcCatalog and examine the results.

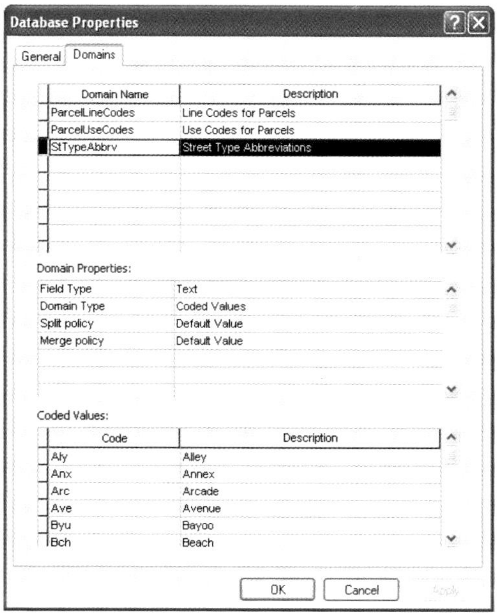

Assign the domains

Now that all the domains are created, they need to be assigned to the fields that they will help control. The tables worksheet shows that these domains will be assigned to two fields in the Parcels feature class and one field in the LotBoundaries feature class.

Tables worksheet

Feature class or table name	Field name	Field type	Alias	Nulls (Y/N)	Default value	Domain name or subtype field (D) or (S)
Parcels	SubName	Text	Subdivision Name	No		
	Blk	Text	Block Designation	No		
	LotNo	Text	Lot Number	No		
	Pre_Type	Text	Prefix Type	Yes		
	Pre_Dir	Text	Prefix Direction	Yes		
	House_Num	Text	House Number	No		
	Street_Name	Text	Street Name	No		
	Street_Type	Text	Street Type	No		(D) StTypeAbbrv
	Suffix_Dir	Text	Suffix Direction	Yes		
	ZIPCODE	LI	ZIP Code	No		
	UseCode	Text	Land Use Code	No		(D) ParcelUseCodes
	Georeference	Text	Georeference	No		
	PlatStatus	SI	Plat Status	No	1	(S) PlatSubtype
LotBoundaries	LineCode	Text	Line Code	No		(D) ParcelLineCodes

1 In ArcCatalog, open the properties of the Parcels feature class and click the Fields tab. Click the field Street_Type.

2 In the Field Properties pane, click the blank space next to Domain. Use the drop-down arrow and click StTypeAbbrv. Click Apply. Notice that when you select the domain, the drop-down list is by name and not description. This is another reason to give domains very explicit names instead of something less descriptive or generic like "Material" or "Type."

3 Next, click the field UseCode, move down to the Domain box, and use the drop-down arrow to click ParcelUseCodes. Click Apply, and then click OK to close Feature Class Properties.

> ### YOUR TURN
>
> Apply the domain ParcelLineCodes to the LineCode field in the LotBoundaries feature class.

Set up subtypes

It was recognized earlier that there would be a benefit to having subtypes in the Parcels feature class. A subtype can show the plat status and allow updates to be done more efficiently. If you are still unsure of the role of subtypes, the next section will demonstrate how they can be used to good effect.

Refer to the subtypes worksheet for the information that will be included in the subtype entry screen.

Subtypes worksheet				PRESET DEFAULTS	
Subtype name	Code	Description	Field	Domain name	Default value
PlatSubtype	1	Platted Property			
	2	Unplatted Property			
	3	Plat Pending			

1 In ArcCatalog, right-click the new feature class Parcels and select Properties. Click the Subtypes tab. Click the drop-down arrow for Subtype Field and select PlatStatus.

2 Move down to the Subtypes input area. Replace code 0 with a **1** and type the description of **Platted Property**.

3 Fill in the remaining entries according to the subtypes worksheet:

- 2 = Unplatted Property

- 3 = Plat Pending

When your dialog box matches the image to the right, click Apply, and then click OK.

In this example, there were no additional default or domain values to set up unique to the subtype setting. Pressing Apply was a good way to check if there were any problems with your entries. If an error message had been returned, it would have instructed you how to correct the error and proceed.

Define the relationship class

You designed a relationship class earlier that will link the parcel polygons to an external table of ownership data. This relationship class will be built in the geodatabase for land records, and it can only access data stored within this geodatabase. Review the design form for relationship classes before proceeding.

1 Import the table TaxRecords2010 from the install folder (for example, C:\ESRIPress \GIST3\Data\) into your new LandRecords geodatabase by right-clicking the geodatabase name, pointing to Import, and clicking Table (single).

2 Set Input Rows to TaxRecords2010 and Output Table name to the same, saving it in your MyAnswers folder. Click OK.

3 Start the creation of the relationship class by right-clicking the LandRecords geodatabase, selecting New, and then clicking Relationship Class.

4　Using the information from the geodatabase worksheets, enter the name of the relationship class as **Ownership**, select Parcels as the origin feature class, and TaxRecords2010 as the destination table. Click Next.

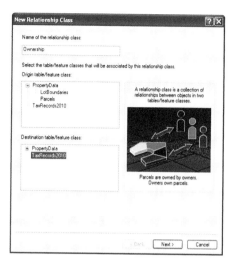

5　Set the relationship type to Simple (peer to peer). Click Next.

6　Type the labels that will describe the two relationships—**Parcel is owned by** for the forward path (as you traverse from the origin table), and **Owner has ownership of** for the backward path (as you traverse from the destination table). Set the message condition to None. Click Next.

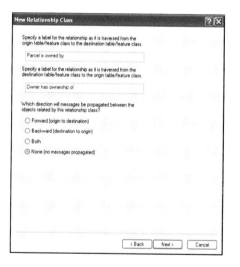

7　Set the relationship cardinality to M-N (many to many). Click Next.

8　Select the option indicating that you want to have attributes for this relationship. Click Next.

9　Type the new field name **PercentOwn** with a data type of Float, as recorded on the design form. Set Field Properties to allow for null values. Click Next.

10 Use the two drop-down lists to set the origin and destination primary key fields to Georeference and GeoReferen, respectively. Then type the labels for the foreign key fields from the design form (see tutorial 1–1)—**Property** for the origin table and **Owner** for the destination table. Click Next.

11 The last screen is a summary of the values you have entered. Review them for completeness, and then click Finish. The new relationship class has now been created in your geodatabase. This links the feature class Parcels with the table TaxRecords2010 so that the information from the table can be used for analysis and symbolized through the parcels. As new tax rolls are released, more relationship classes can be built allowing for comparisons of different tax years to be displayed in the Parcels feature class.

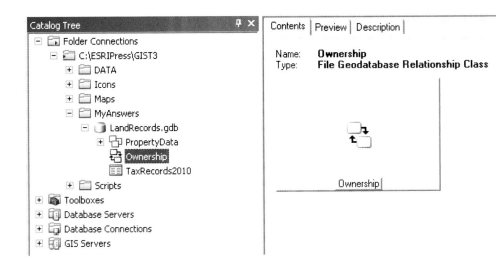

12 Close ArcCatalog.

Test the subtypes

To get a better idea of the value of subtypes and the other data integrity rules, it would be beneficial to test them in ArcMap.

1   Start an ArcMap session and open Tutorial 2.1.mxd.

2–1

2–2

2   Click the Add Data button, browse to the LandRecords geodatabase, click the PropertyData feature dataset, and then click Add. When a feature dataset is added to a map document, all the feature classes in the feature dataset are added. Since these feature classes are empty, no features will be displayed in the map area. However, the table of contents will show some interesting things.

The two feature classes Lot Boundaries and Property Ownership now appear in the table of contents. Normally when a feature class is added to a map document, it is symbolized with a single symbol. In the case of a feature class with a subtype, however, the data is automatically classified according to the subtype field. Notice that the Property Ownership layer shows the three categories defined as subtypes.

3   From the Editor toolbar, select Start Editing. When the edit session is started, ArcMap creates a new editing menu called the Create Features template. All the feature classes will appear in this window shown with their classifications. To edit, simply click the feature you wish to work with, select an editing tool, and then begin working in the map area.

4   In the Create Features template, select Platted Property.

5 Try drawing a few polygons in the map area, and then click the Attributes button on the Editor toolbar. Notice that only the Plat Status field is populated.

6 Click the empty box next to the Land Use Code field. Notice the drop-down list. Its values come from its domain. Also look at the domain for Street Type. With Platted Property selected in the Create Features template, all new features drawn will automatically be put in the Platted Property subtype and all the unique data integrity rules you set up will be in effect. Without subtypes, you would have to go into the attribute table and change each entry as it is drawn.

The next tutorial will build more feature classes with data integrity rules. These will be tested to give you an even better idea of the role of subtypes and domains.

7 Close ArcMap without saving your edits or the map document.

Exercise 2–1

The tutorial showed how to create a geodatabase and all its components from the logical model. This included a feature dataset, feature classes, domains, and subtypes.

In this exercise, you will repeat the process using the logical model for zoning data created in exercise 1–1.

2–1

2–2

- Start ArcCatalog and locate the MyAnswers folder.

- Use the geodatabase worksheet to create the geodatabase.

- Use the tables worksheet to create the feature classes with the correct fields.

- Use the domains worksheet to create and apply the domains.

- Use the subtypes worksheet to create the subtypes.

WHAT TO TURN IN

If you are working in a classroom setting with an instructor, you may be required to submit the materials you created in tutorial 2–1.

Screen capture of ArcCatalog showing all the components of the completed geodatabases

Tutorial 2–1
Exercise 2–1

Tutorial 2–1 review

Any model of reality is only as good as the foundation on which it rests. This tutorial illustrated the development of a geodatabase using a systematic approach to design that documents the uses and relationships of the geographic features of parcels in the City of Oleander.

The tutorials in this book deal mainly with the file geodatabase, but there are other types of geodatabases that you should be aware of. The **file geodatabase** was introduced in ArcGIS 9.2 and is stored as a system of folders containing files. Perhaps the biggest advantages to a file geodatabase are that it is cross-platform compatible; the available file sizes can extend up to one terabyte; and the database can be processed faster. The **personal geodatabase** uses a Microsoft Access database, and consequently, tables can be read in Microsoft Access or Excel. Personal geodatabases are typically used in smaller organizations, where versioning is not required; there are fewer users of the data; and file sizes do not exceed two gigabytes. One disadvantage to the personal geodatabase is that because it is based on Microsoft Access, it can only be used in a Microsoft environment. The **SDE**

geodatabase is cross-platform compatible and can be operated using a variety of DBMS formats. (Check ArcGIS Desktop Help for a current list of supported formats.) Typically used in larger organizations, an SDE geodatabase can support many editors editing the geodatabase simultaneously and efficiently manage these edits through versioning. The SDE geodatabase requires ESRI ArcSDE software to help translate geoprocessing commands to the database. The logical model that you design can be used to build any of these types of geodatabases.

Through the use of **geodatabase design forms** to design your geodatabase, specific representations and necessary relational behavior among geographic features can be clearly represented. The use of the design forms provides a **logical model** of the geodatabase and focuses the developer's planning throughout the development process to ensure adequate representation of these features inside the geodatabase. As such, the development of a logical model is a central aspect of any geodatabase planning phase.

ArcCatalog was used to organize data in **feature datasets**, **feature classes**, and **subtypes** within the geodatabase. You can picture the hierarchical relationships of the geodatabase, feature datasets, and feature classes as a filing cabinet, where the geodatabase is the drawer of the filing cabinet, the feature dataset is the folder that resides in the drawer, and the feature classes are the individual pieces of paper within the file folder. You can assume that an organized relationship exists among these components in a filing cabinet if you ever want to find a document filed away. You can also assume that documents that are similar to each other, or that must support each other, are filed in the same folders. Remember, the rules of the geodatabase require that all feature classes supporting a topology or network must exist within the same feature dataset.

Data integrity and consistency issues within the LotBoundaries and Parcels feature classes of the geodatabase were addressed through the development of **domains** for several features. Proper use of domains in the beginning of design can save hundreds of hours of data cleanup from mistakes made during the data entry phase and countless hours of interpreting nonstandardized entries of data. Although domains are specifically used by feature classes, they are established at the geodatabase level to ensure that one central place will contain all domains and be available for all feature classes. Once a domain is created for a field within a feature class, the likelihood of making mistakes when entering data into this field is reduced drastically, thus improving data integrity and consistency issues within the geodatabase.

During the initial planning phase of geodatabase development, you should examine the relationships among all feature classes and the features they contain with respect to their common behaviors and characteristics. Typically, you want to optimize (minimize) the number of feature classes contained in your geodatabase, while still capturing all the unique characteristics of each feature. This opportunity for optimizing your geodatabase is one of the advantages of using **subtypes** within a feature class. Creating subtypes for a feature class based on certain defining characteristics of features within the feature class essentially allows you to build in the flexibility of an additional feature class without having to build and store one within your geodatabase. The City of Oleander used subtypes to classify the parcels it manages according to the current status of the plat. The status of a parcel with respect to platting is critical to supporting the city's development activities.

You could have created a separate feature class for each plat status. You would simply create a feature class for unplatted parcels, a separate feature class for platted parcels, and a third feature class for plat pending parcels. However, by recognizing that platted, plat pending, and unplatted characteristics of a parcel are simply a specific state of the evolution of the parcel in the land development process, you created a geodatabase with a subtype that would differentiate each parcel on the basis of its status. All three of these subtypes are still types of parcels. However, now they have been categorized in the geodatabase by a subtype. Regardless of what features you intend to store in a geodatabase, as long as the primary behaviors among features are consistent, considering a subtype instead of creating a separate feature class is a good choice.

2–1

2–2

This tutorial used ArcCatalog to develop subtypes for the Parcels and LotBoundaries feature classes and illustrated another advantage of using subtypes when presenting your data to the user. When you displayed your Parcels feature class in ArcMap, each feature class was automatically classified by these subtypes. Again, rather than adding three separate feature classes to display information on plat status within Oleander, ArcMap used the subtypes from this one feature class to automatically classify features based on subtype. This is a more efficient way of working with and viewing the data. As you will see in later tutorials, the creation of subtypes gives you additional advantages when editing your data. Subtypes created within a feature class will allow each subtype to be a separate target feature during an edit session. This gives the user the ability to use unique default values for each subtype and select unique values for each subtype from the supporting domain values. As you can see, a little thinking ahead regarding the use of subtypes and domains can help not only organize the behavior and characteristics of the features contained in your geodatabase, but also provide many advantages when displaying, editing, and populating the geodatabase with data.

STUDY QUESTIONS

1. List three advantages of using domains in your geodatabase. At what level within the geodatabase are these domains established, and why are they created at this level?

2. Why was it necessary to create separate feature classes for the parcels and the lot boundaries? Could this characteristic have been represented by a subtype?

3. How does the establishment of default values within the table of a feature class benefit the operation of the geodatabase?

4. How can importing domains for a feature class benefit the geodatabase developer? Provide an example of suitable information that could be imported into a domain.

Other real-world examples

The use of domains and subtypes to refine the management of spatial data within a geodatabase can be beneficial to anyone needing to develop a geodatabase that accurately portrays geographic features and their associated behaviors. State and local agencies use domains for roadways to help standardize data entry and enforce data integrity for specific types of paving materials, which could include a choice between asphalt and concrete. Water utilities use domains to help standardize entries for fields that contain pipe size and pipe materials.

Subtypes can be used as a data management tool to help refine the classification of features. A roadway feature class, for example, can be organized to optimize maintenance planning according to the type of material. Subtypes can be used to create a default classification of the roads based on asphalt or concrete construction that would be important to help determine different maintenance programs for each type of road. Default values and domains can then be tailored to the unique demands of each roadway subtype. Using subtypes helps to create more definition in the features without having to create additional feature classes.

Power utilities can create subtypes within the geodatabase for different types of power poles—whether metal or wood. Although each type of pole has the same primary behavior of supporting power lines, transformers, and so forth, they vary by material and capacity, and consequently these subtypes will have implications for maintenance and planning.

Tutorial 2–2

Adding complex geodatabase components

The logical model included several techniques for simplifying data entry and managing data integrity. These are included in the geodatabase when it is built in ArcCatalog. If time is taken to design it well in the beginning, the long-term benefits will be great.

Learning objectives

- *Work with ArcCatalog*
- *Create a geodatabase*
- *Build a database schema*

Introduction

Tutorial and exercise 2–1 worked with polygon feature classes. These have unique characteristics in their database design. The use of domains and subtypes helped build in certain data integrity rules.

This tutorial will also focus on setting up feature classes with domains and subtypes but will work with linear and point data types. The process of setting up and using them is very similar to the polygons used previously, so the tutorial will move rather quickly. When everything is set up, however, a more extensive demonstration of the purpose of the data integrity rules will be presented.

Adding to the data structure

Scenario   You've completed designing and building the database for the parcel and zoning data for the City of Oleander. Now it's time to create the geodatabase structure for the sewer line designs that were also written earlier.

Data   Print out the design forms from tutorial 1–2. These form the completed geodatabase logical model from which the sewer line schema can be created.

Tools used   ArcCatalog:

> New Personal Geodatabase
> New Feature Dataset
> New Feature Class
> Create a Domain

Review the printed geodatabase design forms from tutorial 1–2. There are many components to build, and this tutorial will run through them rather quickly. If you encounter any problems, review the steps from tutorial 2–1 or look in ArcMap Help.

Geodatabase design forms		
Geodatabase name	Utility Data	
Feature dataset name	Wastewater	
Feature classes:		
Type Feature class name	Alias	
L SewerLines	Sewer Lines	
L Interceptors	Interceptors	
PNT SewerFixtures	Sewer Fixtures	
PNT InterceptorFix	Interceptor Fixtures	

Type: Indicate if this is a PoINT, Line, or POLYgon feature class.
Name: Enter the name of your feature class.
Alias: Describe the contents of the feature class.

Create the data structure

The first part of the creation process will be to create the geodatabase and feature dataset. Check the design forms for the correct names.

1   **Start ArcCatalog. In the MyAnswers folder, create a new file geodatabase called Utility Data. Within this geodatabase, begin creating a feature dataset called Wastewater, but stop on the second input screen. This step is asking for the spatial reference of the data to be stored here. It will be the same as the spatial reference for the PropertyData feature class in the LandRecords geodatabase, so it would be faster and more efficient to borrow a copy from that location.**

2   **Click Import and browse to the PropertyData feature class in the LandRecords geodatabase. Once the feature dataset is selected, click Add. Then accept the remaining defaults and finish the creation process.** The Import button is a shortcut for setting the spatial reference. There are other situations where parameters can be imported from existing items, so always be careful to note which dialog screen is active. When the Spatial Reference dialog box is open, the Import button will import a spatial reference. If a different dialog box were open, those parameters would be imported.

3 Go to the Wastewater feature dataset and start the creation process for the SewerLines feature class, being very careful to set the feature type correctly. In the field name dialog box, enter all the fields from the design form. Pay close attention to the Allow Nulls setting and any default values.

Tables worksheet

Feature class or table name	Field name	Field type	Alias	Nulls (Y/N)	Default value	Domain name or subtype field (D) or (S)
SewerLines	PipeSize	SI	Pipe Size	No		(D) SewerPipeSize
	Material	SI	Pipe Material	No		(S) SewerLineMaterial
	YearBuilt	LI	Year Built	Yes		
	Description	Text	Owner	No		

Remember that the domains and subtypes cannot be assigned yet, because they have not yet been created. In the next feature class, the fields are the same as in the SewerLines feature class. There is an import feature that will make filling out the fields dialog box very easy.

Import field definitions

The next feature class on the design form is for interceptors.

Tables worksheet

Feature class or table name	Field name	Field type	Alias	Nulls (Y/N)	Default value	Domain name or subtype field (D) or (S)
Interceptors	PipeSize	SI	Pipe Size	No		
	Material	Text	Pipe Material	No		
	YearBuilt	LI	Year Built	Yes		
	Description	Text	Owner	No		

1 Go to the Wastewater feature dataset again and begin creating the Interceptors feature class. Enter the name and description from the design form and click Next twice. Instead of typing in all the field names, click the Import button to retrieve the fields from another file.

2 Browse to the SewerLines feature class, select it, and click Add.

3 Consulting the design forms, you will see that the field type for Material is Text. Change the field type from Short Integer to Text. Check that the alias and null settings are correct, and then click Finish to complete the feature class. Once again, the Import button was a great shortcut. You saw, however, that one of the field types was different from the imported data structure. This can be changed while you are working in the feature class creation screens, but once you apply the changes, they cannot be changed again. It is interesting to note that field structures can be imported.

Continue the creation process

Now move on to the feature class for the sewer fixtures.

Tables worksheet						
Feature class or table name	Field name	Field type	Alias	Nulls (Y/N)	Default value	Domain name or subtype field (D) or (S)
SewerFixtures	FixType	SI	Fixture Type	No		(S) SewerFixType
	FlowLine	Float	Flow Line Elevation	Yes		
	RimElev	Float	Rim Elevation	Yes		
	Depth	Float	Depth from Surface	Yes		
	YearBuilt	LI	Year Built	Yes	2010	
	Description	Text	Owner	No	Oleander	

1 In the Wastewater feature dataset, create the SewerFixtures feature class. Add the fields and their parameters from the design worksheet. Properties of a field include its data type, alias, nulls flag, default values, and domains. When everything is correct, click Finish.

The final feature class to create is for the interceptor fixtures. Notice that it has four of the same fields as the sewer fixtures.

Tables worksheet

Feature class or table name	Field name	Field type	Alias	Nulls (Y/N)	Default value	Domain name or subtype field (D) or (S)
InterceptorFix	FlowLine	Float	Flow Line Elevation	Yes		
	RimElev	Float	Rim Elevation	Yes		
	Depth	Float	Depth from Surface	Yes		
	YearBuilt	LI	Year Built	Yes	2010	

2 Once again, go to the Wastewater feature dataset and create a new feature class for InterceptorFix. In the Field Name dialog box, click Import and transfer all the fields from the feature class SewerFixtures.

2–1

2–2

3 Two of the fields are not necessary. Remove them by clicking the gray box to the left of their name and pressing Delete. Once they are both removed, click Finish. Now is a good time to go back and check your work. Did you create all the feature classes as the correct feature types? Did you get all the Allow Null Values settings correct? Did you spell all the field names correctly? If not, now is the time to delete the feature class and start again as none of these things can be changed. Also, check field alias names and required default value settings. If any of these are incorrect, they can be changed now.

Create domains

Move on to the domains worksheet. There is only one to create. Notice that the field type is Short Integer and the codes are integers. But the descriptions are text with inch marks included. Both of these formats will be saved in the domain, and both are available for use in a map document. For example, when a label is created to show pipe size, you will have the choice of using 12 or 12".

Domains worksheet

Domain name	Description	Field type	Domain type	Coded values / Range Code (Min)	Desc (Max)
SewerPipeSize	Sewer Pipe Size	SI	Coded values	6	6"
				8	8"
				10	10"
				12	12"

1 Open the properties of the Utility Data geodatabase and click the Domains tab. On the first blank line, enter the name of the domain and its description and adjust the Domain Properties. Proceed to the Coded Values pane and enter the other parameters and codes for the domain. Click Apply, and then click OK. This pipe size domain will be assigned to the PipeSize field in the SewerLines feature class.

2 Open the properties of the SewerLines feature class and click the Fields tab. Click the PipeSize field and in the Field Properties pane, use the Domain drop-down arrow to assign the SewerPipeSize domain. Click Apply, and then click OK.

Create subtypes

The last of the data integrity rules involve the subtypes. Review the subtypes worksheet.

Subtypes worksheet

Subtype name	Code	Description	Field	Domain name	Default value
				PRESET DEFAULTS	
SewerLineMaterial	1	P.V.C.	PipeSize	SewerPipeSize	8
			Description		Oleander
			YearBuilt		2010
	2	HDPE	PipeSize	SewerPipeSize	10
			Description		Oleander
			YearBuilt		2010
	3	DI	PipeSize	SewerPipeSize	12
			Description		Oleander
			YearBuilt		2010
	4	Conc			
	5	Clay			
SewerFixType	1	Manhole			
	2	Cleanout			

The second subtype is less complex, so for learning purposes, you should create that one first. Then a more detailed description will follow to create the more complex subtype.

1 Right-click the SewerFixtures feature class, click Properties, and then click the Subtypes tab. Use the drop-down arrow to set the subtype field to FixType.

2 Change the first code from 0 to **1** and enter the description **Manhole**. Add a second code of **2** with a description of **Cleanout**. Click Apply, and then click OK.

Now to tackle the tougher subtype. This one has codes and descriptions like the other subtype but also includes a domain and default setting for each subtype code.

3 Right-click the SewerLines feature class, click Properties, and then click the Subtypes tab. Use the drop-down arrow to set the subtype field to Material.

4 Change the first code from 0 to **1** with a description of **P.V.C.** Then in the Default Values and Domains pane, enter the default values of **8** for PipeSize, **2010** for YearBuilt, and **Oleander** for Description. Click Apply to check for errors. If there are any, correct them before moving on.

5 On the next blank line, enter a code of **2** with a description of **HDPE**. Enter the default values of **10** for PipeSize, **2010** for YearBuilt, and **Oleander** for Description. Click Apply.

6 Continue by entering a code of **3** and a description of **DI** on the next blank line. Enter the default values of **12** for PipeSize, **2010** for YearBuilt, and **Oleander** for Description. Click Apply.

7 The last two subtypes do not have any additional default or domain parameters. On the next two lines, enter codes **4** and **5** with descriptions of **Conc** and **Clay**, respectively. Click Apply, and then click OK. This wraps up the database creation process. All the components of the logical model have been entered. On very complex datasets, it's a good idea to run back through the logical model and check the entries against the feature classes to make sure nothing was missed. If you were methodical in the process, everything should be OK.

8 Close ArcCatalog.

Test the rules

As in the last tutorial, it would be advisable to test the features in ArcMap before proceeding. This quick pilot study will determine if any of the rules are not working and will give you the chance to go back into the data structure and correct them if necessary. Once data is loaded into the feature classes, it would be much riskier to try and fix problems, especially if this involved deleting and re-creating a field.

1 Start ArcMap and open Tutorial 2–2.mxd. Add the feature dataset Wastewater from the Utility Data geodatabase to the table of contents, and zoom to the bookmark Starting Extent. All the feature classes in the Wastewater feature dataset should be added to the table of contents. Notice that the layers Sewer Fixtures and Sewer Lines have already been classified by their subtype values. ArcMap does this automatically, although if this is not the desired classification, it can be changed manually.

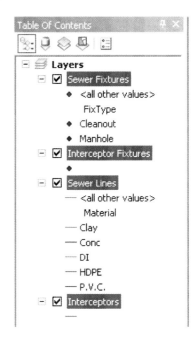

2 Go to Editor > Start Editing. The Create
Features template will display all the
features that can be edited, grouped by
feature class and subtype. Select P.V.C.
from the Sewer Lines listing. Notice that
the list of selections under Sewer Fixtures
and Sewer Lines comes straight from the
subtypes you defined earlier.

3 Use the default construction tool in the
map area to draw a sample line, complete it by double-clicking the last point,
and open the Attributes dialog box from
the Editor menu. Notice in the Attributes
dialog box that all the fields have already
been filled in! These are the default values
that you defined in the layer properties. For
the sewer line type of P.V.C., the default
size is 8", the year built is 2010, and the
description regarding ownership of the line
is set to Oleander.

YOUR TURN

Draw each of the types of sewer lines and notice the attribute values assigned to each one.
Remember that the defaults apply only to new items drawn. Since concrete and clay pipes are
rarely installed as new, there are no defaults set for these. They need to appear in the database,
however, to accommodate existing pipes made of these materials. If they were not in this list,
they would test as invalid when the database is checked.

4 **When you have finished looking at the various sewer lines, close ArcMap without saving edits or the map document.** This was a quick test of the data structure. If any errors were found, you could go back and correct them in ArcCatalog, but be sure to close ArcMap first before making any corrections. If you wish to delete incorrect entries from the Attribute table, start an edit session in ArcMap, highlight the bad entries, and press Delete. Remember that once data is added to these feature classes, it will be more difficult to change the data structure after the fact.

Exercise 2–2

The tutorial showed how to create a geodatabase and all its components from the logical model. This included a feature dataset, feature classes, domains, and subtypes.

In this exercise, you will repeat the process using the logical model for a storm drain system created in exercise 1–2.

- Start ArcCatalog and locate the MyAnswers folder.
- Use the geodatabase worksheet to create the geodatabase.
- Use the tables worksheet to create the feature classes with the correct fields.
- Use the domains worksheet to create and apply the domains.
- Use the subtypes worksheet to create the subtypes.

WHAT TO TURN IN

If you are working in a classroom setting with an instructor, you may be required to submit the materials you created in tutorial 2–2.

Screen capture of ArcCatalog showing all the components of the completed geodatabases

Tutorial 2–2
Exercise 2–2

Tutorial 2–2 review

This exercise built on the concepts of creating unique **attribute domains** and **subtypes** for linear features. The City of Oleander is managing its sewer network in a geodatabase. As you can see, the same process that was used for setting up domains and subtypes for parcels was used for the sewer system. By using the standardized planning process and filling out the forms ahead of time, you were able to simplify creation of the geodatabase for the sewer system.

2–1

2–2

ArcCatalog provides all the necessary tools to easily create your geodatabase; however, careful thought was necessary to help establish the behaviors and relationships for these features. As you could see, separate feature classes could have been created for the different materials of pipes and fixtures. However, regardless of the material of pipe or fixture, each one has the same basic function and could be managed more efficiently through the creation of **subtypes** based on material. Using subtypes to define categories of features in this manner is common in the industry to distinguish behavior, attributes, and access properties of features. A domain was created for the size of the sewer pipes to help with the standardization of these values in order to enforce data integrity.

As you have worked through the tutorials, you have probably noticed that there has always been a step to test each component that you built into your geodatabase to ensure that it is representing the features as the needs of the project dictate. This QA/QC exercise is a critical step that should be conducted as often as possible throughout the development project. QA/QC of your database development project should not be skipped because of time constraints, or for any other reason. There is nothing worse than completing the construction of a project and realizing that it does not adequately meet the needs of your client because of a logical oversight or because an error was made early in the development process. Typically, these situations can surface in two different ways that you can help control with thorough planning: Either the features, attributes, behaviors, access restrictions, and so forth, built into the model do not adequately support the workflow process they were designed to support; or errors were made in the execution of the plan. It's really pretty simple. When a failure occurs, you either have a bad model or poor execution. (Of course, this is assuming that the workflow process itself is designed adequately to achieve the necessary results. Most likely, you will have no control of the process that you are hired to support).

To help you succeed with your database design project, make sure to do the following: (1) meet with and document your interviews with project sponsors to discern the necessary requirements for the project, (2) have the project sponsor review and approve the project requirements document that you compiled, (3) use the geodatabase design forms to carefully think through an appropriate approach that reflects the approved results from your requirements session, (4) review your design with the project sponsor and get his or her approval of your proposed design, (5) carefully execute your plan of developing your geodatabase, (6) conduct testing to ensure that your geodatabase meets your requirements, and finally (7) implement your geodatabase.

Many of the fastest ways to fail in a GIS database development project can be avoided simply by carefully checking your work, having a colleague check your work, and then rechecking your own work one final time. One of the advantages of using the design forms during the planning phase is that you create a documented paper trail of your logic that you can refer to if a failure occurs and you need to troubleshoot your design. The ESRI geodatabase provides many of the tools you will need to help build an efficient, flexible, accurate, and consistent model of reality. In the last two chapters, you have learned how to organize your geographic features in a geodatabase and how to incorporate attribute domains and feature subtypes into the geodatabase. These tools, as well as the many others that you will be learning in upcoming tutorials, will help you become successful, but remember that the most critical component of any GIS is *the capable mind of a thinking operator!*

STUDY QUESTIONS

1. Explain the rationale for creating subtypes for the sewer lines. What other approach could have been used and how? What situations prohibit the use of subtypes? What advantages do subtypes offer during data editing?

2. What are the primary advantages of creating domains for the sewer pipes? How will this step save you time later?

3. Explain the relationship between feature datasets and feature classes. What are the advantages and disadvantages of grouping your feature classes into several feature datatsets?

4. List three mistakes that could be made when creating a feature class that would cause you to have to re-create the entire feature class. How can these mistakes be avoided?

5. What is the most critical component of any GIS?

Other real-world examples

Domains and subtypes are a handy way to increase efficiency when populating your geodatabase and are used in a variety of situations. Consider the use of subtypes and domains in the case of infrastructure management at an electric utility. An electric utility manages its customer power distribution network through a set of power generators, wires, transformers, power poles, and meters. Subtypes can be used to categorize different types of power poles. Wooden power poles can support up to three transformers, and metal power poles can support up to five. The power poles essentially perform the same function in supporting the transformers and the wires; however, when data is being entered, a wooden power pole domain is created to limit the transformer load to three transformers, and a metal power pole domain is created to limit the transformer load to five transformers. Simply establishing these domains to restrict data entry values to established rules for each subclass of power pole will prevent inaccurate data from being entered into the geodatabase, and will allow each type of pole to be symbolized using different symbols on a map (see chapter 8, "Developing labels and annotation").

Another example of the use of domains and subtypes can be found in the water utility industry. Consider valves along the water transmission pipeline, for example. Valves are used to regulate water within the pipelines. There are two primary types of water valves: maintenance valves and control valves. Maintenance valves are used primarily in pipeline maintenance activities and water flow efficiency regulation. Control valves can be used to restrict or prevent water from flowing within the pipe system. Subtypes can be created for each valve type to designate a maintenance or control function. Within each of these valve subtypes, there are different varieties of valves such as blow-off or air valves in the maintenance valve subtype, and butterfly or gate valves in the control valve subtype. Domains can be created for each subtype to ensure that maintenance valve attribute fields are populated only with blow-off or air valve values and that control valves are populated strictly with butterfly or gate valve types. Such a process will not only streamline data entry, but also enhance data entry accuracy.

References for further study

There are a number of additional resources to help create a geodatabase. **ArcGIS Desktop Help** lets you use keywords to search for a title or topic. You can access ArcGIS Desktop Help in ArcMap or ArcCatalog by clicking Help on the main menu. Then click the Search tab, type a keyword, and click Ask. Use the keywords provided here to search for additional learning resources on many of the concepts taught in this chapter.

ESRI has developed a number of online courses on a wide variety of pertinent GIS topics. **ESRI Self-Study (Virtual Campus) courses** are an excellent resource for students that will supplement information covered in this course. ESRI Virtual Campus courses can be accessed at `http://training.esri.com/gateway/index.cfm`.

ArcGIS Desktop Help search keywords

geodatabase, attribute domains, feature class, feature datasets, working with geodatabases, building geodatabases, feature class basics, subtypes, data integrity, topology, projections, spatial reference, coded values, creating subtypes

ESRI Self-Study (Virtual Campus) courses

The following ESRI Virtual Campus classes may be helpful:

1. Basics of the Geodatabase Data Model
2. Creating, Editing, and Managing Geodatabases for ArcGIS Desktop
3. Working with Geodatabase Subtypes and Domains
4. Creating and Editing Geodatabase Topology with ArcGIS Desktop

3

Populating a geodatabase

Tutorial 3–1

Loading data into a geodatabase

There are lots of data sources out there, and any one of them can be brought into a geodatabase. This first look at importing data will deal with bringing data from a polygon shapefile into a geodatabase.

Learning objectives

- *Import data*
- *Work with shapefile data*
- *Create a load procedure*
- *Load data into subtypes*

Introduction

There are lots of ways to put data into a geodatabase format. The simplest is to find the data in ArcCatalog, right-click it, and select Export > To Geodatabase. With this process, however, you have no control over the data structure of the output file. The file will have the same data structure as the input data and won't have the built-in data integrity rules that you designed into the data in the last few tutorials.

A better process would be to load data carefully into the new geodatabase structure, taking account of the different feature classes and their subtypes. This is done by right-clicking one of the new empty feature classes you created and selecting Load Data. Then the Simple Data Loader guides you through the process, letting you map fields and populate the subtype structure. It's the difference between dumping a wheelbarrow full of data into your computer versus making your own containers and carefully selecting what goes into each one. It is important to note that any mistakes in loading the data cannot be undone, so all the imported data would have to be deleted and the process started over. Thus, you should be very careful to set each parameter correctly for each load procedure.

The load process can also be tedious, requiring that you repeat the process many times to separate the data into a new framework. Imagine one big dataset that contains all the water, sewer, and storm drain utilities. Your new design might split these into four or five feature classes, each with a number of subtypes. You would need to run the load process for each feature class–subtype combination, which could mean that the process is run twenty or thirty times before all the data is incorporated into the new data structure. But once it is completed, the new data integrity rules and other functions built into the geodatabase structure will make editing and updating much more streamlined.

Establishing a load procedure

Scenario   Having gone through the design process for the parcel data and come up with a pretty good design, you can now populate the new geodatabase and start using the new format. After digging through the archives, city staff discovered some old shapefiles containing the data you need. You will use the Simple Data Loader to bring the old data into the new data structure.

Data   You already have the empty containers ready—the LandRecords geodatabase created in the previous tutorials. The existing shapefile with the parcel data is called ParcelSource.shp and is located in the C:\ESRIPress\GIST3\Data folder.

The shapefile contains a field called PlatStatus, which will be used to match the data to the subtypes:

1 = Platted Property

2 = Unplatted Property

3 = Plat Pending

The other fields for ParcelSource.shp, which will be used to match the source fields to the new geodatabase, are as follows:

Prop_Des_1	subdivision or abstract name
Prop_Des_2	lot and block designation
Acreage	area of parcel in acres
DU	number of dwelling units on parcel
PlatStatus	status of the platting process
UseCode	land use code
PIDN	tax office identifier
Prefix	street prefix (N,S,E,W)
StName	street name
Suffix	street suffix
SufDir	street suffix direction (N,S,E,W)
LotNo	lot number
BlkNo	block number
PID1	subdivision code
PID2	lot or tract number
PID3	tax transaction code
PID4	tax transaction code
Prop_Add	situs address

3–1

3–2

Addno	address number
EKEY	tax account number
GeoReferen	tax office georeference key
YearBuilt	year of construction
Marker	internal use
TADMap	tax office map number
PoliceDist	police district number
RandomNum	random parcel selection
HouseValue	property appraised value
CityOwned	city-owned property
Shape_Leng	perimeter of parcel (provided by ArcMap)
Shape_Area	area of parcel (provided by ArcMap)

Field Name	Data Type
FID	Object ID
Shape	Geometry
Prop_Des_1	Text
Prop_Des_2	Text
Acreage	Float
DU	Short Integer
PlatStatus	Short Integer
UseCode	Text
PIDN	Text
Prefix	Text
StName	Text
Suffix	Text
SufDir	Text
Lotno	Text
BlkNo	Text
PID1	Text
PID2	Text
PID3	Text
PID4	Text
Prop_Add	Text
Addno	Long Integer
EKEY	Long Integer
GeoReferen	Text
YearBuilt	Short Integer
Marker	Text
TADMap	Text
PoliceDist	Text
RandomNum	Long Integer
HouseValue	Double
CityOwned	Text
Shape_Leng	Double
Shape_Area	Double

Tools used   ArcCatalog:
Simple Data Loader

The process begins with looking at the existing data and deciding which parts will go where in the new data structure. The old data may have more fields than needed, but it may also be missing data that was included in your design. Once a field map is created, the process will become clear.

Create a field map

The field map will be a list of all the fields from your new data structure with their counterparts listed from the source data.

1 Print a copy of the completed tables worksheet from the geodatabase design forms you made for the Parcels layer in tutorial 1–1. On a separate sheet, copy the list of the fields you have designed. Compare these to the aforementioned field names and descriptions for the ParcelSource.shp file. To the right of each field from the Parcels layer, write the name of the source field for that data and try to match the field descriptions. Write your own first, and then check it against the accompanying chart.

SubName	Prop_Des_1
Blk	BlkNo
LotNo	LotNo
Pre_Type	<None>
Pre_Dir	Prefix
House_Num	Addno
Street_Name	StName
Street_Type	Suffix
Suffix_Dir	SufDir
ZIPCODE	<None>
UseCode	UseCode
Georeference	GeoReferen
PlatStatus	PlatStatus

That solves the question of which fields match up as far as data loading, but there are also the subtypes to consider. In the source data, Plat Status is shown by an integer value of 1, 2, or 3. As you load the data, you'll use a query to select each of these codes and load it into the subtype you set up previously.

Start the loading process

1 In ArcCatalog, browse to the location of the new LandRecords geodatabase and PropertyData feature dataset you created in tutorial 2–1. Right-click the feature class Parcels and select Load > Load Data. This will open the Simple Data Loader.

2 Click the Browse button, go to the Data folder, and select ParcelSource.shp as the input data. Click Open.

3–1

3–2

3 Then click Add in the Simple Data Loader dialog box and click Next.

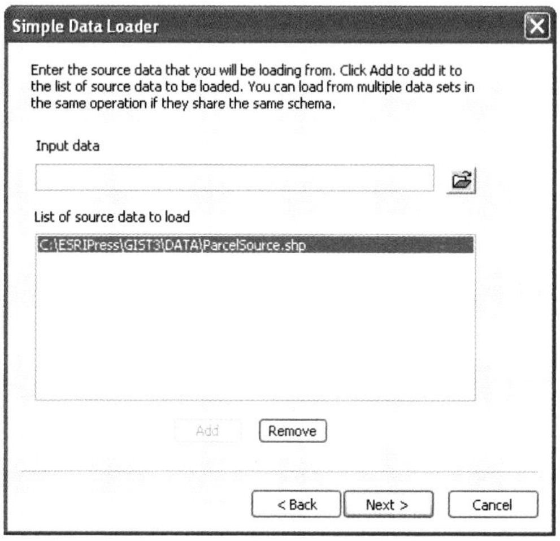

4 The next screen asks for the destination geodatabase and feature class. These were established by right-clicking an existing feature class, so they appear dimmed. At the bottom, click the selection to load into a subtype. Use the drop-down arrow to set the target subtype to Platted Property. Click Next.

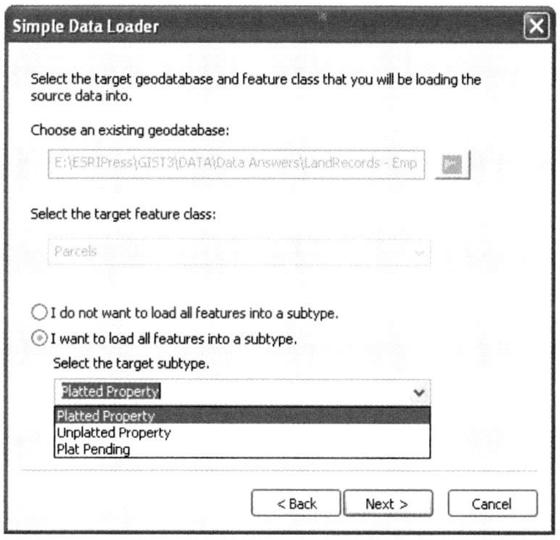

Now is where the field mapping comes into play. The box lists the target fields and asks you to identify the source fields. Use the diagram you made in the first step to fill in the target field matrix.

5 Click the Matching Source Field column in the first row, opposite the field SubName. Use the drop-down arrow and click Prop_Des_1 (string).

Simple Data Loader

For each target field, select the source field that should be loaded into it.

Target Field	Matching Source Field
SubName [string]	<None>
BLK [string]	<None>
LotNo [string]	Prop_Des_1 [string]
Pre_Type [string]	Prop_Des_2 [string]
Pre_Dir [string]	Acreage [float]
House_Num [string]	DU [short int]
Street_Name [string]	PlatStatus [short int]
Street_Type [string]	UseCode [string]
Suffix_Dir [string]	PIDN [string]
ZIPCODE [int]	Prefix [string]
	StName [string]
	Suffix [string]

YOUR TURN

Continue down the list from the field map. Set each source field to match the table and click Next.

Simple Data Loader

For each target field, select the source field that should be loaded into it.

Target Field	Matching Source Field
Pre_Type [string]	<None>
Pre_Dir [string]	Prefix [string]
House_Num [string]	Addno [int]
Street_Name [string]	StName [string]
Street_Type [string]	Suffix [string]
Suffix_Dir [string]	SufDir [string]
ZIPCODE [int]	<None>
UseCODE [string]	UseCode [string]
Georeference [string]	GeoReferen [string]

Reset

< Back   Next >   Cancel

Notice that some of the fields were already correctly mapped. The Simple Data Loader looks for standard field names or those spelled the same in both the target and source datasets and automatically matches them. Be careful, however, because it may match fields that do not contain the same data, such as Prefix and Prefix Direction.

6 The next screen asks for a query to separate the data, or it will load all the data without regard to the subtypes. Click "Load only the features that satisfy a query." Then click Query Builder. Build the query **"Plat Status"** = **1**. Click OK, and then click Next.

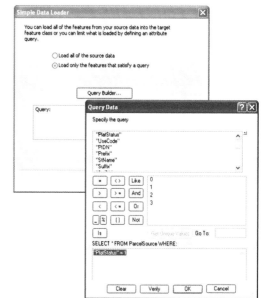

3–1

3–2

7 The final screen is a summary of the process. Examine it to make sure everything looks OK and click Finish.

The load won't take very long. You've created the load procedure that has loaded the first subtype. There are two more to load—the unplatted property and the plat-pending property. Note that the input data has the plat-pending property coded as 0 and 3, so you will have to load both into the proper subtype.

YOUR TURN

Repeat the process with the Simple Data Loader, starting by right-clicking the Parcels feature class in ArcCatalog. Change the selection to load into the Unplatted Property subtype and build the query **"Plat Status"** = **2**. Then go through it once more, changing the subtype to Plat Pending and build the query **"Plat Status"** = **3**.

Check the results in ArcCatalog

1 Click the Preview tab to see the graphics that were imported into your feature class. Zoom and pan around the data and notice that it is color coded by the subtype field Plat Status.

2 Next, change the preview type to Table and notice that all the data was loaded into the correct fields.

SubName	Blk	LotNo
MIDWAY SQUARE ADDITION	I	19
MIDWAY PK	21	21
EULESS MAIN PL	3	4
MIDWAY PK	26	41
MIDWAY PK	23	8
MIDWAY PK	24	31
MIDWAY SQUARE ADDITION	J	18
HIDEAWAY ADDITION	A	22
MIDWAY PK	27	14
MIDWAY PK	21	33
MIDWAY PK	23	31
MIDWAY SQUARE ADDITION	K	6
MIDWAY PK	26	13

Load the linear data

The process of loading the polygons that represent parcels is completed. Next you will load the data that represents the edge of the parcels. The previous tutorial explained that because the edges may need to be symbolized individually, they need to be in their own feature class. You'll also learn later how to manage the association between these features using other data integrity rules.

1 Right-click the LotBoundaries feature class and go to Load > Load Data to start the Simple Data Loader.

2 Set the input data to ParcelLineSource.shp located in the Data folder. Click Add, and then click Next.

3 Keep clicking Next until you get to Finish to start the data transfer. Notice that the LineCode target field was automatically matched with the correct field from the input source data, because the field name was identical except for casing.

4 When the process is complete, preview the geography and table for the LotBoundaries feature class.

| Contents | Preview | Description |

OBJECTID *	SHAPE *	Linecode
612	Polyline	1
613	Polyline	2
614	Polyline	1
615	Polyline	2
616	Polyline	1
617	Polyline	2
618	Polyline	1
619	Polyline	1
620	Polyline	2
621	Polyline	1
622	Polyline	2
623	Polyline	2
624	Polyline	1

3-1

3-2

All the data has been successfully loaded into the geodatabase. This is the location where the data will be stored and maintained for many years to come. If you like, start an ArcMap session and drop the data into an empty map document to see how it looks. The geodatabase has many data integrity rules built into it but retains the flexibility to be modified to accommodate future expansion and analysis.

Exercise 3–1

The tutorial showed how to use the Simple Data Loader to selectively add data to an existing geodatabase–feature class structure.

In this exercise, you will repeat the process and load data into the zoning geodatabase you made in exercise 2–1. This will include both the polygons and lines that represent the zoning districts and their boundaries.

- Start ArcCatalog and go to the ZoningData geodatabase that you created in the MyAnswers folder for exercise 2–1.

- Use the Simple Data Loader to load the zoning data into this geodatabase. The two files containing the source data are in the Data folder. The file ZoningSource.shp contains the polygon information, and the file ZoningLineSource.shp contains the zoning boundary information. Investigate their data structure if necessary to determine the best procedure for importing the data. Remember to import into the subtypes you created.

- When completed, test the data integrity rules in ArcMap.

WHAT TO TURN IN

If you are working in a classroom setting with an instructor, you may be required to submit a screen capture of the geodatabase you loaded in tutorial 3–1.

Screen capture of the geodatabase previews from

        Tutorial 3–1 (geography and table)
        Exercise 3–1 (geography and table)

Tutorial 3–1 review

Data creation is one of the costliest aspects of GIS implementation that organizations face. Consequently, the ability to **import** existing data from other digital formats into your geodatabase has the potential for significant cost savings. Tools contained in ArcMap and ArcCatalog have proven to be very successful in importing existing digital data from other formats into the geodatabase. One of the key reasons that these tools are successful is because of the flexibility they provide to customize the import of geospatial data based on user needs and the geospatial data structure and format. As illustrated in the last two chapters, developing a good plan to import your data is crucial to the success of building your geodatabase.

ArcCatalog offers a variety of import capabilities for existing **shapefiles**. One way to import shapefile data into your geodatabase is to simply let ArcCatalog create a feature class from the existing shapefile features and attributes. While this is the most expedient option in the short term, it may cost you later when you attempt to enhance the functionality, standardization, and security of your geodatabase according to established user requirements. Another way to import data is to use ArcCatalog to import your data into the structured design of the existing geodatabase you created. To accomplish this in the exercise, you created a **load procedure** that instructed ArcCatalog to "match" or "map" the import shapefile data to existing fields and subtypes in the geodatabase. As you have already devoted quite a bit of time to organizing your data according to its feature types, relationships, and desired functionality, you will save time by transferring import data directly into the geodatabase according to the rules established in the load procedure.

By using the **Simple Data Loader** in ArcCatalog and its **field mapping** capabilities to assign each feature and attribute from the import source data to its appropriate feature type and subtype in the target feature class, you were able to make use of your established geodatabase design to import data into the geodatabase. Although this method initially required more time than using an automated means of importing your data into the geodatabase, it will pay for itself in the long run. As in the initial planning activities during the geodatabase design phase, the Simple Data Loader option of importing data was enhanced by a "load procedure" to determine where features and attributes of the import tables will reside in the geodatabase. This process may involve multiple iterations until every feature and subtype is appropriately "matched" to the existing geodatabase structure, but it will pay for itself in the long run.

3–1

3–2

STUDY QUESTIONS

1. Name two options for transferring data from an existing shapefile to a geodatabase. Explain the advantages and disadvantages of each option.

2. Given what you already know about feature subtypes, why is it important to have the ability to import features directly into a subtype in the existing geodatabase?

Other real-world examples

The ESRI shapefile is a common format of geospatial data. As illustrated by the exercise, importing shapefiles can be a pretty straightforward process. However, this is not to suggest that only shapefiles can be exported into the geodatabase. You can load coverage, shapefile, CAD, or geodatabase feature class data into an existing feature class, provided it falls within the spatial reference of the feature class you're loading. Additionally, you can load INFO, dBase, or geodatabase table data into an existing table or append data to an existing feature class or table in your geodatabase if needed. Often in the work environment, you will encounter situations where you have a table of information that you need to append to an existing table in your geodatabase. This can also be achieved by using the Append or Merge geoprocessing tools with the associated field mapping functionality to append data from the export table to the table in your geodatabase.

Another likely situation is to import data in an ArcInfo coverage format. During import of an ArcInfo coverage file, all layers contained in the ArcInfo coverage will create corresponding feature classes in the resulting geodatabase (i.e., points, lines, polygons, or annotation). One thing you must consider when importing ArcInfo coverage features is that the ArcInfo coverage format maintains several feature types that may not be necessary to portray in the geodatabase.

On the other hand, since importing digital data provides an opportunity for tremendous cost savings to an organization, it is important to realize that a variety of digital spatial data can be successfully imported into your geodatabase. There is a wealth of information available regarding data formats and import techniques, so don't be afraid to experiment. As you gain experience importing different data formats into the geodatabase, you will learn what considerations are important for each of the formats.

Tutorial 3-2

Populating geodatabase subtypes

Loading data from various sources is all achieved in the same manner, regardless of what the data represents. The feature types and fields may be different, but they are still loaded into a geodatabase using the Simple Data Loader and carefully crafted queries.

Learning objectives

- *Import data*
- *Work with shapefile data*
- *Create a load procedure*
- *Load data into subtypes*

Introduction

Now you've seen how the load process works. Although it may be tedious, it is the best way to control with precision how your data is loaded into a geodatabase. The linear and point features here will load exactly the same way as the polygon and linear features in the last tutorial. Just be diligent in setting the subtypes and queries to make sure everything loads correctly.

Because this process can be so repetitive and tedious, it is a good candidate for modeling, which can be done in ModelBuilder. This topic will be covered later in the book, so you may want to repeat this tutorial after completing the ModelBuilder tutorials in chapter 7.

Loading into multiple feature classes

Scenario You completed the geodatabase design for the sewer system in chapter 1, created the geodatabase in chapter 2, and now will be populating it with previously created sewer line data that the city had from another mapping endeavor. As before, you'll use the Simple Data Loader to carefully move the data from the shapefile to the geodatabase. This time, however, the data will be loaded into several feature classes from the same data source, and subtypes will be used to segregate the data even further. Be very careful to include all the parameters and queries you'll need to use with the Simple Data Loader to ensure success. Any misloaded data will be cause to start all over again.

Data   You already have the empty containers ready—the Utility Data geodatabase created in
the previous tutorials. The existing shapefiles with the sewer system data are called
SewerLineSource.shp and SewerNodeSource.shp. They are located in the Data folder.

The SewerLineSource.shp contains the linear data concerning the pipes. It will be loaded
into subtypes in the new feature class using the material type. It will also be split between
two feature classes, one for city-owned pipe and one for interceptors that are owned by
a regional treatment facility. The following field descriptions will be used to match the
existing data to the new field structure:

PSize: Pipe size

Material: Pipe material (P.V.C., HDPE, DI, Conc, Clay)

Year_Const: Year of construction

Comments: Notation of pipe's function

      Oleander = standard collection pipe

      Interceptor = large-size interceptor pipe

The SewerNodeSource.shp contains the data for the manholes and cleanouts, as referenced
in the designs. Once again, be aware of the subtypes, and note that the data will be split
into two feature classes just as the pipes were.

SYMB: Code for manholes or cleanouts

      11 = manhole

      15 = cleanout

Depth: Depth of line from surface

Fline: Flow line

Year_Const: Year of construction

Comments: Notation of fixture's function

      Oleander = standard collection fixture

      Interceptor = large-size interceptor fixture

RimElev: Rim elevation

Tools used   ArcCatalog:
              Simple Data Loader

As before, it is a good idea to make a field map, or a diagram that shows which fields in the
new feature class will be matched with which fields from the source data.

Create a field map

The field map will be a list of all the fields from your new data structure with their counterparts listed from the source data.

1 Print a copy of the completed tables worksheet from the geodatabase design forms you made for the SewerLines layer in tutorial 1–2. On a separate sheet, copy the list of the fields you have designed. Compare

PipeSize	Psize
Material	Material
YearBuilt	Year_Const
Description	Comments

these to the aforementioned field names and descriptions for the SewerLineSource.shp file. To the right of each field from the SewerLines layer, write the name of the source field for that data and try to match the field description. Write your own first, and then check it against the table shown.

3–1
3–2

Load the data

Remember also that the data will be split into the feature classes SewerLines and Interceptors, based on the contents of the Comments field. This will be done with a query in the Simple Data Loader.

1 Start ArcCatalog and browse to the Utility Data geodatabase you built in tutorial 2–2. Go to the Wastewater feature dataset, right-click the SewerLines feature class, and start the Simple Data Loader.

2 Set the input data to SewerLineSource.shp located in the Data folder. Click Next.

3 Click "I want to load all features into a subtype," set the target subtype to P.V.C., and click Next.

4 Use the chart you made in the first step to set the field mapping. Click Next. Notice that the Material field doesn't appear in the field mapping done by the Simple Data Loader. That is because it is the subtype field, and it is being populated automatically as a result of your previously selecting the "Load into a subtype" option.

In the next step, you will set the data loader to load only data that belongs in the P.V.C.

subtype, as well as the pipes that belong to Oleander. This can be done with a complex query. Try writing the query first, and then check it against the next step.

5 **Click "Load only the features that satisfy a query," and then click Query Builder. Build the query:**

"MATERIAL" = 'P.V.C.' AND "Comments" = 'Oleander'

6 **Click OK, and then click Next. Review the summary and click Finish when you are satisfied that everything looks OK.** This has loaded all the pipes that belong to Oleander and that are made of P.V.C. There are several more material types to load, so the process needs to be repeated for each additional type.

YOUR TURN

Run the Simple Data Loader to load each of the following material types into the SewerLines feature class. Remember to also set the query to only load pipes owned by Oleander.

HDPE (HDPE), Clay (V.C.)

There are additional subtypes of R.C.C.P. for Reinforced Concrete Pipe and DI for Ductile Iron, but there are no existing pipes of these materials, so you do not need to run the Simple Data Loader for these subtypes. **Note:** If you make a mistake in loading the data, there is no way to undo the action. So, be careful.

When you are done, click the Preview tab to see the results. Examine both the geography and the table. If completed correctly, you will notice that the lines are different colors to represent the subtypes.

PipeSize	Material	YearBuilt	Description
8"	P.V.C.	2002	Oleander
8"	P.V.C.	2002	Oleander
8"	P.V.C.	2002	Oleander
8"	P.V.C.	2002	Oleander
8"	P.V.C.	2002	Oleander
8"	P.V.C.	2002	Oleander
8"	P.V.C.	2002	Oleander
8"	P.V.C.	2002	Oleander
8"	P.V.C.	2002	Oleander
8"	P.V.C.	2002	Oleander
8"	P.V.C.	2002	Oleander
8"	P.V.C.	2002	Oleander

Now, you have loaded all the features for the SewerLines feature class, but there are still features in the source data that need to be handled. That is the data for the interceptors.

7 Browse to the Utility Data geodatabase you built in tutorial 2–2. Go to the Wastewater feature dataset, right-click the Interceptors feature class, and start the Simple Data Loader.

8 Set the input data to SewerLineSource.shp located in the Data folder. All the interceptor data will go into one feature class without being divided into subtypes. Because of this, the field Material will also appear on the field mapping screen.

3–1

3–2

9 On the next screen, click "I do not want to load all features into a subtype." Then click Next.

10 Set up the field mapping as before. The Material field is automatically matched, because the field names are the same for the source and the target. When your screen matches the graphic shown here, click Next. Remember that the source file holds all the pipe data, and just as you did with the Oleander pipes, you will need to query only the interceptors for this feature class.

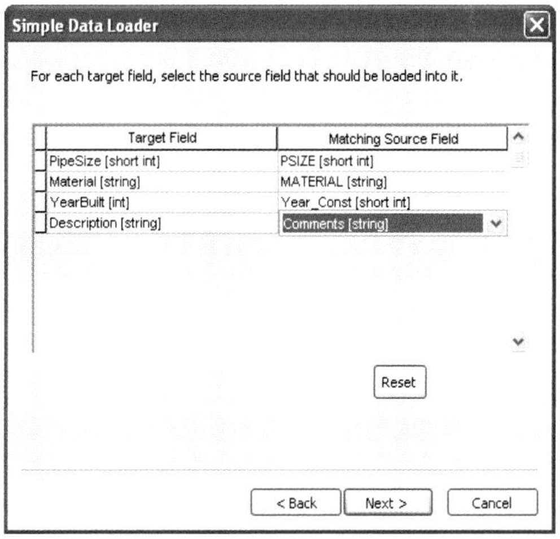

Simple Data Loader

For each target field, select the source field that should be loaded into it.

Target Field	Matching Source Field
PipeSize [short int]	PSIZE [short int]
Material [string]	MATERIAL [string]
YearBuilt [int]	Year_Const [short int]
Description [string]	Comments [string]

Reset

< Back   Next >   Cancel

11 Click "Load only the features that satisfy a query," and then click Query Builder. Build the following query:

"Comments" ='Interceptor'

12 Click OK, and then click Next. Review the summary and click Finish when you are satisfied that everything looks OK.

Examine the results

1 Preview the geography and table of the Interceptors feature class. It should show just the larger lines that were selected with the query.

OBJECTID *	SHAPE *	PipeSize	Material
1	Polyline	15	V C
2	Polyline	15	V C
3	Polyline	15	V C
4	Polyline	15	V C
5	Polyline	15	V C
6	Polyline	15	V C
7	Polyline	15	V C
8	Polyline	15	V C
9	Polyline	24	R C C P
10	Polyline	15	V C
11	Polyline	15	V C
12	Polyline	15	V C
13	Polyline	42	R C C P
14	Polyline	24	R C C P
15	Polyline	24	R C C P
16	Polyline	42	R C C P
17	Polyline	42	R C C P
18	Polyline	42	R C C P
19	Polyline	42	R C C P
20	Polyline	42	R C C P
21	Polyline	15	V C
22	Polyline	42	R C C P
23	Polyline	24	R C C P
24	Polyline	45	R C C P
25	Polyline	45	R C C P
26	Polyline	45	R C C P
27	Polyline	24	R C C P

Load the point data

The process should be pretty familiar by now, since it has a built-in repetitive nature. Start by making a field map to determine what fields from the source data will match the new SewerFixtures feature class in the Wastewater feature dataset. The data description at the start of this tutorial will help.

1 Print a copy of the completed tables worksheet from the geodatabase design forms you made for the SewerFixtures layer in tutorial 1–2. On a separate sheet, copy the list of the fields you have designed. Compare these to the aforementioned field names and descriptions for the SewerNodeSource.shp file. To the right of

FixType	SYMB
FlowLine	Fline
RimElev	RimElev
Depth	Depth
YearBuilt	Year_Const
Description	Comments

each field from the SewerFixtures layer, write the name of the source field for that data and try to match the field descriptions. Write your own first, and then check it against the accompanying chart.

The next task will be to load the point data representing the sewer line fixtures into the SewerFixtures feature class. This will again involve loading the data into individual subtypes based on the SYMB field. Check back to the data descriptions at the start of this chapter to see which values will be used.

2 Right-click the SewerFixtures feature class in the Wastewater feature dataset where you've been working. Start the Simple Data Loader. Set the input data source to SewerNodeSource.shp in the C:\ESRIPress\GIST3\Data folder. Add it to the list, and click Next to continue.

3 Set the Simple Data Loader to load data into the Manhole subtype.

4 Set the field mapping to match the diagram you made earlier.

5 Use the Query Builder to select and load only the Oleander manhole data.

6 Review the settings and click Finish to complete. When the process is finished, you should have only the manhole features loaded into the SewerFixtures feature class.

YOUR TURN

There's one more subtype to load: cleanouts. Start the Simple Data Loader and load only the cleanouts for Oleander into the SewerFixtures feature class.

This should now bring the total features to 1,948 and should be all the Oleander data. You're closing in on the finish line with only one more task to complete: The interceptor fixtures will all go into the InterceptorFix feature class without any subtype divisions.

YOUR TURN

Find the InterceptorFix feature class in the area where you've been working, start the Simple Data Loader, and load the interceptor fixtures from the SewerNodeSource shapefile. Use the query **"Comments"** = **'Interceptor'** to find the desired features.

Check the results

1 Preview the geography and table for the SewerFixtures feature class.

OBJECTID *	SHAPE *	FixType	FlowLine	RimElev
1448	Point	Manhole	510.06	520.31
1449	Point	Manhole	508.27	518.17
1450	Point	Manhole	501.4	519.75
1451	Point	Manhole	211.2	519.38
1452	Point	Manhole	523.58	533.39
1453	Point	Manhole	522.77	533.2
1454	Point	Manhole	524.86	532.86
1455	Point	Manhole	525	530.95
1456	Point	Manhole	514.77	527.97
1457	Point	Manhole	596.48	601.95
1458	Point	Manhole	602	607.12
1459	Point	Manhole	595.8	601.35
1460	Point	Manhole	0	0
1461	Point	Manhole	506.46	519.25
1462	Point	Manhole	507.54	519.2
1463	Point	Manhole	510.37	520.2
1464	Point	Manhole	0	0
1465	Point	Cleanout	0	0
1466	Point	Cleanout	0	0
1467	Point	Cleanout	0	0

2   Now preview the geography and table for the InterceptorFix feature class.

The import process is finally complete. As advertised, it's very repetitive and tedious.

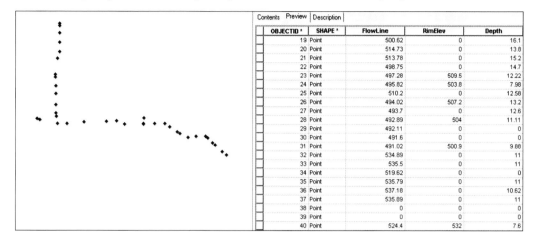

It is important to pay attention to the minute details of the process in order to get through it cleanly and correctly. Otherwise, you'll have to delete what you've loaded into the geodatabase and start all over.

Exercise 3–2

The tutorial showed how to use the Simple Data Loader to selectively add data to an existing geodatabase.

In this exercise, you will repeat the process and load data into the storm drain geodatabase you created in exercise 2–2. This will include both the lines and points that represent the storm drain pipes and their fixtures.

- Start ArcCatalog and browse to the StormDrain geodatabase in the MyAnswers folder.
- Investigate the fields in the Storm_Fix_Source.shp file in the Data folder and match them to the fields in your new polygon feature class. Repeat with the Storm_Line_Source.shp file and the new linear feature class.
- Make a list of the fields in your destination feature classes and match them to the fields in the source subtypes before you start the data loading.
- Use the Simple Data Loader to load the storm drain data into the subtypes.

- The fixture codes are as follows:

  106 = "Y" inlet

  110 = beehive inlet

  101 = curb inlet

  102 = grate inlet

  109 = headwall

  104 = junction box

  107 = junction box/manhole

  111 = manhole (personnel access port)

  108 = outfall

Hint: Since these are already stored in an integer field in the source data, you will not have to load them individually into the subtypes. Once they are loaded, ArcCatalog will see these as correct subtype values.

- When completed, test the data integrity rules in ArcMap.

WHAT TO TURN IN

If you are working in a classroom setting with an instructor, you may be required to submit a screen capture of the geodatabase you loaded in tutorial 3–2.

Screen capture of the geodatabase previews from

> Tutorial 3–2 (geography and table)
> Exercise 3–2 (geography and table)

Tutorial 3–2 review

The **Simple Data Loader** available in ArcCatalog provides an easy, guided approach to importing geographic features into the geodatabase. As you can see from exercise 3–2, importing point and line features into the geodatabase uses the same process that you completed for polygons in exercise 3–1. Again, a well-thought-out approach to importing features from a shapefile into your geodatabase includes developing worksheets that document all your decisions during the import of your data.

Although the **shapefile** containing the sewer line data included all the different types of sewer lines in one shapefile, the import process allowed you to separate the sewer lines into two different feature classes according to the owner of the sewer line. To complete this most effectively, you used the **query functionality** within the Simple Data Loader to query the Comments data field and appropriately isolate each of the sewer lines to load your data into new feature classes for sewer lines and interceptor lines. The import process also allowed you to further segregate the data you imported into existing **subtypes** in the geodatabase by populating the sewer line subtypes with the appropriate material for each pipe. For the points data, it let you delineate whether the subtype of each sewer node was a manhole or a cleanout.

One thing that should be evident by now is the importance of understanding your data. Familiarity with the data that is being imported as well as knowledge of how that data will be represented and manipulated inside the geodatabase is critical to your success. Exercises 3–1 and 3–2 provided examples of a real-life scenario where data was imported from shapefiles that each stored data differently. By using the **Simple Data Loader** tools in ArcCatalog and your **load procedure**, you were able to import these datasets in an organized and systematic fashion. Successful importing of the data required knowledge not only of how it was stored and managed in the shapefile, but also of how you wanted the data to be stored and managed in your geodatabase. In the real world, you will have to conduct the same logical steps to properly load source data into a target geodatabase.

3–1

3–2

STUDY QUESTIONS

1. What is a load procedure, and why is it important for you to develop and follow one? What knowledge is required to successfully develop and execute the load procedure?

2. Using the Simple Data Loader, what option must you specify before objects are loaded into a subtype?

3. What is the best way to quickly determine whether the import procedure was correctly performed?

Other real-world examples

This tutorial focused on the effective process of importing line and point spatial data into a geodatabase using the Simple Data Loader tool in ArcCatalog. It is critical that the process is done correctly to ensure successful implementation. However, successful implementation also requires that you increase your knowledge of the data being imported into the geodatabase. Two major areas to consider involve the (1) accuracy and suitability of the data being imported into the geodatabase, and (2) the manner in which this data was used by the organization in its original format and how the geodatabase will enhance its use.

Your knowledge of the accuracy and suitability of the data can be enhanced by understanding how the data was originally collected. Was the sewer line data collected through GPS field collection efforts? Were the locations of the manholes transcribed and digitized from field notes? Were the parcel lines digitized from existing paper plans? How current is the data? These questions will contribute to your knowledge regarding the accuracy and quality (and ultimate success) of the data being imported into your geodatabase. If you imported bad data correctly, you have only perpetuated the problem within the geodatabase. This is not to suggest that only data of the highest quality may be imported, because regardless of the quality, it may be the only representations you have of the features to be included. Rather, it is simply to ensure that the analyst notes the quality of the data in the metadata of the feature classes that are built into the geodatabase.

The second major area contributing to your success concerns methodology and investigating such questions as "How was the data used in the original file?" and "How did the data in the imported file satisfy user requirements?" In the example regarding sewer lines, all the sewer lines, regardless of ownership, were contained in one shapefile. By determining how this data would be used and managed, you were able to determine in the initial geodatabase design process that the organization would benefit from having the sewer lines characterized by ownership and managed in separate feature classes. By gaining a thorough understanding of such workflow issues prior to the import process, you will gain insight into developing load procedures. A concurrent knowledge of where the data comes from, with your plan and procedures to import the data, and where you want to go with it will increase your likelihood of success when developing and populating your geodatabase.

References for further study

There are a number of additional resources to help you load data into your geodatabase. **ArcGIS Desktop Help** lets you use keywords to search for a title or topic. You can access ArcGIS Desktop Help in ArcMap or ArcCatalog by clicking Help on the main menu. Then click the Search tab, type a keyword, and click Ask. Use the keywords provided here to search for additional learning resources on many of the concepts taught in this chapter.

ESRI has developed a number of online courses on a wide variety of pertinent GIS topics. **ESRI Self-Study (Virtual Campus) courses** are an excellent resource for students that will supplement information covered in this course. ESRI Virtual Campus courses can be accessed at `http://training.esri.com/gateway/index.cfm`.

ArcGIS Desktop Help search keywords

importing data, shapefile, loading data, Simple Data Loader, shapefile extensions, mapping fields, query, query subtypes, geodatabase, attribute domains, feature class, feature datasets, working

ESRI Self-Study (Virtual Campus) courses

The following ESRI Virtual Campus classes may be helpful:

1. Creating, Editing, and Managing Geodatabases for ArcGIS Desktop
2. Working with Geodatabase Subtypes and Domains
3. Creating and Editing Geodatabase Topology with ArcGIS Desktop (for ArcEditor and ArcInfo)
4. Creating and Maintaining Metadata Using ArcGIS Desktop

4

Working with features

Tutorial 4–1

Creating new features

There are many techniques for creating new line features in ArcGIS, including the Editor toolbar, Advanced Editing tools, and COGO tools. With these tools, any level of accurate data can be drawn in ArcMap, whether it comes from sketched diagrams or survey-quality plans. The key is in matching the tools to the task.

Learning objectives

- *Set up group layers*
- *Set up basemap layers*
- *Organize the table of contents*
- *Work with bookmarks*
- *Set map extents*
- *Create new features*
- *Use the context menu tools*

Introduction

The last tutorial showed how to take existing data and import it into a carefully crafted geodatabase structure. But not all the data you might need will have already been created. There comes a time when you will have to create the data yourself. ArcMap contains a large array of tools for creating points, lines, and polygons, and the trick to creating new features is to determine which tool to use. Sometimes this is done with absolute coordinates, sometimes with survey data, and sometimes by describing a feature's location in relation to other features.

This tutorial will use a combination of tools to turn a rough schematic into GIS data.

Sketching in the details

Scenario   The Oleander Regional Transportation Authority (ORTA) has received the first draft of plans for a new rail line and commuter station to be located in Oleander. It has provided a measured drawing, but unfortunately, it is not based on clean survey data. Instead, it is a collection of data from which you will need to try and reproduce the project as an overlay for the existing Oleander data.

The city planner would like to see the train track alignment overlaid on the parcels to study the potential impact of noise and pollution. The planner would also like to see the location of the station and parking lots to determine any potential traffic problems.

The drawing provided should be enough to sketch in the proposed location with adequate detail to do the preliminary study.

Data   Empty line and polygon feature classes are provided in the ORTA geodatabase, which you may use to draw the plan. Use ArcCatalog to copy the ORTA geodatabase from the Data folder to the MyAnswers folder. Also provided is the Oleander parcel layer named Property Ownership created in tutorial 3–1.

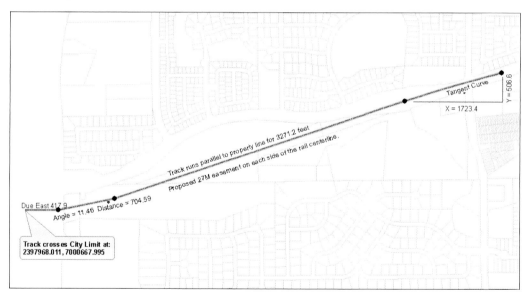

Tools used   ArcMap:

Group layers	Line: Direction/Length
Bookmarks	Line: Parallel
Specify Extent	Line: Length
Line: Absolute X, Y	Tangent tool
Line: Delta X, Y	Fillet tool

Take a look at the diagram provided showing the proposed rail line and station. The first step will be to draw the rail line. The data seems pretty haphazard and disjointed, but with the right techniques, the rail line can be accurately drawn. This will involve using a combination of the standard editing tools and the context menu tools.

Organize the table of contents

1　**Start ArcMap and open Tutorial 4–1.mxd.** In this map document, the table of contents includes several layers. The map displays all the parcels and property boundaries located within the city of Oleander, the survey control monuments, and some empty feature classes that will contain the new data for the proposed rail facility.

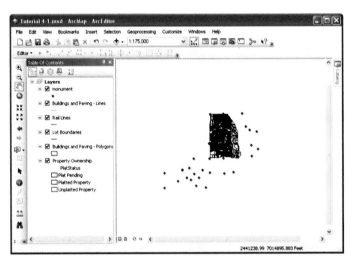

There's a lot going on in this map document, and none of it looks good. Before you can begin to edit and add new features, you will need to do some housekeeping to make the map document easier to work with. The first task will be to tidy up the table of contents. The feature classes aren't in any particular order, and the names need to be more descriptive of the project. Remember that when a legend is added to the map document, the names of the layers appear and need to describe the project. As a management tool, change the name of the data frame to be more reflective of the project, add two new group layers to the table of contents, and then arrange the layers into these.

2　**Right-click the data frame name—the word Layers at the top of the table of contents. Click Properties, and then click the General tab. Type the new name of Proposed Rail Station. Click OK.** An easier and faster way to rename any data frame or layer is to click the name, pause slightly, move the mouse a tiny bit, and then click again. The name will be in Windows edit mode, and you will be able to retype it. Be careful not to click too rapidly or it will be detected as a double-click and open Layer Properties. It might take some practice to get the hang of it.

Next you will add a special type of group layer to hold the basemap information. This will contain the layers that are not being edited in this map document, making them easier to manage, but it will also store them in a special display cache that will enhance their drawing rate when panning.

3   **Right-click the data frame name again and click New Basemap Layer.** A basemap layer is a device to hold feature classes and allows common properties to be set for all the feature classes it contains. It also lets you control the visibility of all the layers it contains with a single mouse click. Since basemap and group layers look and act just like feature classes, it's advisable to always put the word Group in their name. These groups will also be shown in the legend, so they are very useful in keeping similar layers together for the benefit of the viewer of the printed map output. However, items placed in a special basemap group will not appear in the editing dialog box when an edit session is started and will not be available for snapping while editing.

4   **Change the name of New Basemap Layer to Property Base Group. You may decide whether to open the Properties dialog box and change the name, or try the Windows edit mode.** This new group needs to contain the following layers:

- Property Ownership

- monument (because this is existing data with no alias given, it has come in with a lowercase "m")

These can be moved around simply by dragging the layers into the group. A horizontal bar will be drawn to show where the layer will display, and when the bar is indented, it indicates that the layer will drop into a group.

5   **Drag the Property Ownership layer into the Property Base Group. Watch for the horizontal bar to indent under the group name.**

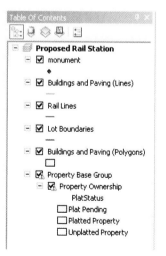

6 **Drag the monument layer into the Property Base Group as well.** The monument layer needs a better name. When this appears in the legend, the viewer should be able to understand what data that layer contains, and right now the layer name is too simple and nondescript.

7 Change the name of the monument layer to **Survey Control Points**. Use either the **Windows editor shortcut, or change it in Layer Properties.** All this makes a huge difference in how the table of contents looks. But there are other benefits to the basemap layer.

8 **Right-click the Property Base Group, and then click Properties. Next, click the General tab.** Under this tab, you can change the name and visibility of the group, add a description or data credits, and set a visible scale range. These parameters will be applied to all the layers in the group.

The visible scale ranges of the layers and the basemap layer can be manipulated to give very precise control over everything in the group. For example, the group may contain three layers of data that are only appropriate at medium zoom scales, one layer of specific data that should only be shown at a close zoom, and annotation that should also only be shown when zoomed in tight. You could set an individual scale range for the two layers that require a tight zoom, and then set an overall scale range for the group that would turn off the group when zoomed out too far. This matrix of visible scale range settings will work together to turn on data only when appropriate.

9 **Click the Group tab to examine more parameters. Note that layers can be added or removed from the group by using this dialog box. A limited list of layer properties can be set as well.** The Group tab allows you to add more layers to the group, remove layers from the group, set the properties of any of the group members, and change the display order. All these things can be done via the table of contents controls but are done here much easier with the click of a single button. At the bottom is an advanced feature called Symbol Levels, which will be discussed later in this book.

10 Finally, click the Display tab for more parameters. When you are done examining the parameters, click OK. This dialog box lets you set the contrast, brightness, and transparency all at once for the entire group. The first two controls are only applicable to raster layers, but transparency will affect all the layers. Be careful if you have set a transparency level for any one of the layers previously, because this setting will reduce the transparency level even more by the percentage you set. ArcMap will simulate the transparency levels in the legend, making the output map easier to read and understand when transparencies are used.

YOUR TURN

Add a standard group layer for the North Oleander Station data. Add the layers for buildings and rail lines to the new group. Leave the Lot Boundaries layer outside the group.

The group layers are a very useful way not only to arrange the data, but also to control some of the common properties across the layer list. Groups can also be nested, so that a group containing basemap information might contain a group for Oleander data, a group for Tarrant County data, and a group for State of Texas data. This will result in a very clean, organized table of contents, and as a result, a very attractive legend. The symbol levels, which will be used later, can control the drawing order even across groupings.

Work with bookmarks

The map is almost ready to begin editing, but first the area of interest must be located. The designers have previously zoomed to the area of interest in order to study the site, and they bookmarked the extent. In ArcMap, bookmarks can be saved and passed around to other users, and the designers have saved one for you.

1 On the main menu, go to Bookmarks > Manage.

2 In the Bookmarks dialog box, click Load and browse to the Data\Bookmarks folder. Click ORTA_SITE.dat, and then click Open.

3 This previously saved bookmark appears in the dialog box. Click it, and then click Zoom To. Then close the Bookmarks Manager dialog box. The map is now zoomed to the area of interest, and the table of contents is tidy with the layers named appropriately.

It's almost time to start editing, but first try panning the data to see the effect of the special basemap group you made.

4 Select the Pan tool from the Tools toolbar and slowly pan the data to the north. Notice that the layers in the basemap group layer redraw as you go.

Control default map extents

The last thing to change is the zoom extents of the Zoom to Full Extent button. When editing, you will be zooming to small areas to see more detail, and then wanting to zoom out to the full extent of the study area to see the results. While bookmarks are very useful for this, the Zoom to Full Extent button is much faster and more convenient. First, try out the button and see the results.

1 **Click Zoom to Full Extent on the Tools toolbar.** The map zooms out to a view seemingly from "outer space." No work could be done with any precision at this zoom level.

Return to the previous zoom level and set this as the new default zoom extent.

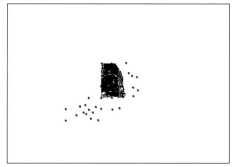

2 On the Bookmarks menu, click ORTA Site to return to the area of interest.

3 Next, open the properties of the data frame Proposed Rail Station and click the Data Frame tab.

4 In the Extent Used by Full Extent Command pane, click Other.

5 Click Specify Extent, and then click Current Visible Extent. Click OK, and then OK again to close Properties.

6 **Test this by panning to a new area, and then clicking Full Extent on the Tools toolbar.** Setting the extents this way is an extremely useful way to prevent the outer space view of your data from constantly appearing. By default, the extent of the full zoom is set to the largest extent of the layers in your map document. If you happen to have brought in some state or national data, your default zoom extent will be very large. Now you have better control over this and will save a lot of time traveling to and from outer space.

This seemed like a lot of work just to set up the map document, but it will pay dividends in the future in the following ways:

- You'll save time by not getting lost in uncontrolled zooms.
- The data is easier to see and understand in group layers.
- The special basemap layer will help speed up panning as well as keep unnecessary layers off the Editing menu.
- You'll save time when you add a legend, because the data is already named and organized.
- You'll be able to turn whole sets of data on and off by the group layers.
- You can use the group Layer Properties dialog box to manage the layer properties of the group.
- Anyone opening this map document will find it easier to understand the data.

Create line features

Now back to the editing tasks.

Begin by examining the alignment of the rail line, which will be drawn as a single line representing the centerline of the track. It begins at a known x,y coordinate and describes the location through various dimensions. These dimensions were taken from field observations, known rail standards, and other methodologies. Although it would have been easier if a single method had been used, it doesn't always happen that way in the real world.

To draw this line, you will use a variety of tools from the Editing context menu. By combining several of them, you'll be able to draw the line. For a complete list of all the context tools available, search for About Creating New Features in ArcGIS Desktop Help.

1 **Make sure that the Editor toolbar is visible, and then begin the creation of the proposed rail line facility by clicking Editor > Start Editing.** Because the data in this map document comes from multiple geodatabases, you must define which geodatabase you wish to edit. Remember that ArcMap can only edit one geodatabase at a time. The upper pane of the dialog box will show all the editable layers from your table of contents. Notice that the layers in the special basemap layer are not shown since they cannot be edited. As you select the geodatabases in the lower pane, the icons for the layer will change to include a pencil to show which layers would be editable if that geodatabase was selected.

2 **Click the ORTA.mdb geodatabase. Notice that the icons for the buildings and rail layers change. You may want to expand the display area of the geodatabase names so that you can read the entire name. Click OK to start an edit session.** The edit session will open the Create Features editing template, displaying the editable layers. To create new features, you will select the desired feature from this dialog box and use the construction tools to draw it in. Note that the layers you placed in the basemap layer group do not appear in the Create Features template, nor are they open for editing.

3 In the Create Features template, click Rail Lines, and then click Line in the Construction Tools pane. Notice that the layer names give a clear idea of what would be edited with each selection. Imagine if the Create Features template were to have a list such as Layer 1, New Features, Lines, and Old Polygons. Would you know which one to select? You can appreciate the value of giving your layers very descriptive names.

The first tool that you will use to create the proposed rail line is the Absolute X, Y tool. On the diagram, the x- and y-coordinates for the beginning of the rail line are provided. These are in unprojected map coordinates, which the Absolute X, Y tool will convert to a location on your map.

4 In the display area of the map document, right-click to expose the context menu and select the Absolute X, Y tool.

5 Enter the coordinates **2397968.011** for the x-value and **7000667.995** for the y-value, as was shown in the diagram. Be careful not to press Enter until both values are correct. Move between the two entry boxes by clicking the mouse or using the Tab key and press Enter when both values are correct. This will enter the first vertex of the new line, and you will see a red dot showing the location in addition to the rubber band stretching to the cursor location. As you use more of the Sketch Construction tool functions, more vertices will be added to the line.

The next tool to use will be the Delta X, Y, which uses the change in the x-direction and the change in the y-direction to locate the next vertex.

6 Right-click the display area again, and from the context menu, select the Delta X, Y tool.

7 Enter **417.9** for the x-value and **0** for the y-value. Remember to use the Tab key to move between the entries and press Enter when both are correct. With these two points entered, the line will start to form as the edit sketch. This is a temporary construction, shown in green, and will not become a feature in the target layer until you double-click to end it, or select Finish Sketch from the context menu. The last location entered is always shown in red, with the rest of the locations and the line you are constructing shown in green.

Route Measure Editing	▶
~~Insert Vertex~~	
Delete Vertex	
Move…	
Move To…	
Change Segment	▶
Flip	
Trim to Length…	
Part	▶
Delete Sketch	Ctrl+Delete
Finish Sketch	F2
Finish Part	
Sketch Properties	

By the way, if you make a mistake when entering a node in your line, move the cursor over the node, right-click to expose the context menu, and click Delete Vertex.

Snap To Feature	▶
Direction…	Ctrl+A
Deflection…	
Length…	Ctrl+L
Change Length	
Absolute X, Y…	F6
Delta X, Y…	Ctrl+D
Direction/Length…	Ctrl+G
Parallel	Ctrl+P
Perpendicular	Ctrl+E
Segment Deflection…	F7
Replace Sketch	
Tangent Curve…	
~~Find Text~~	~~Ctrl+F~~
Streaming	F8
Delete Sketch	Ctrl+Delete
Finish Sketch	F2
~~Finish Sketch~~	
Finish Part	

Or for an easier way to delete an incorrect entry, you could also click the Undo button. Notice on the context menu that many of the tools have a shortcut key, such as F6 for Absolute X, Y and Ctrl+D for Delta X, Y. It's a great time saver if you can remember the ones you use the most.

Next, use the Direction/Length tool. This will accept the angle of the line along with the distance to locate the next vertex. Note its keyboard shortcut for future use.

8 Expose the context menu and click Direction/Length (or press Ctrl+G). Enter the direction **11.46** and the length **704.59**. This adds another vertex and extends the line.

Direction/Length	☒
11.46	704.59

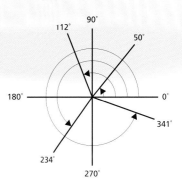

The direction value given in the sketch is in a polar angular measurement. This means that angles start at a single pole, or zero position, and are measured counterclockwise in degrees.

To make sure that your ArcMap session is set up to use polar measurement, open the Options dialog box from the Editor menu and make the appropriate settings.

4–1

4–2

4–3

4–4

4–5

The next segment will be constructed using a combination of tools. The first will be a context menu selection to make the edit sketch parallel to an existing feature. The second will be a different context menu selection to set the length. Once the angle and distance values are given, ArcMap has enough information to construct the next segment and will extend the edit sketch.

In order for the Parallel function to work, you have to identify for ArcMap which feature you wish to get the angle from. This is done by placing the edit sketch cursor, shown as a blue circle with a crosshair, over the desired line and invoking the Parallel function from the context menu. Another way to activate the Parallel function is from the ghost toolbar called Feature Construction. It appears in your map area in a translucent form until you move the cursor over it. It contains several of the more commonly used editing functions and can save time in not having to move the cursor back and forth to the main menu or constantly having to right-click to expose the context menu.

There is an existing property boundary that represents the edge of the railroad right-of-way that will be used to set the angle of the new line.

9 On the Feature Construction toolbar, select the Constrain Parallel tool.

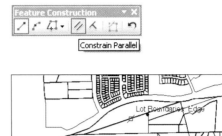

10 Move the cursor over the property line you wish to be parallel to, as shown in the accompanying image. The line will be identified as Lot Boundaries: Edge by the SnapTip. Click the lot boundary.

Note that you could also set the sketch angle by right-clicking the line, and then clicking Parallel, or by hovering the cursor over the lot line and pressing Ctrl+P. Notice that if you move the cursor, the angle of the edit sketch will remain parallel to the line you chose. If you had selected the wrong line, or did not want to use that angle, you could release this constraint by pressing Escape.

11 Right-click and select Length from the context menu (or press Ctrl+L). Type the length as **3271.2**. This portion of the edit sketch will be drawn and you will be ready for the next location.

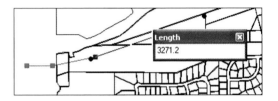

The final segment of the rail line curves off to the edge of the area of interest. This will require a different Sketch Construction tool called the Tangent Curve Segment tool to draw circles.

There are nine Sketch Construction tools, as shown in the figure, and more information can be found in ArcMap Desktop Help by searching for "Sketch Construction tools."

The Tangent Curve Segment tool will draw an arc at the tangent angle from the last sketch segment drawn. With this tool and a length, you will be able to finish the construction.

12 On the Feature Construction toolbar, click the Sketch Construction tool drop-down arrow and select the Tangent Curve Segment tool.

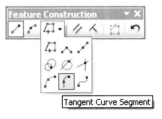

13 Expose the context menu and click Delta X, Y, or press Ctrl+D. Type the values of **1723.4** for x and **506.6** for y. Finally, select Finish Sketch from the context menu or press F2 to complete the creation of the new feature. Be sure to save your edits periodically to avoid losing your work. The line draws out OK according to the sketch, but the angular nature of the turn from the first segment you entered to the second looks odd.

Railroads should curve and flow better than that. It's time to invoke a little artistic license and draw a curve there to make things look better.

14 If you don't already have it, add the Advanced Editing toolbar to your map document and select the Fillet tool. The purpose of the Fillet tool is to round out the angle between two lines, either by eye in sketch mode or by using a detailed dimension.

15 Click the line on one side of the angle where the curve is to be drawn, and then click the line on the other side to specify where you want to construct the fillet. The SnapTip will help you identify the lines.

16 By dragging the cursor, you can control the radius of the fillet, or curve. Drag it out to what you feel is a nice-looking curve and click to add the fillet. This line creation seems like a long process, but as you become more familiar with the Sketch Construction tools, things will go much more smoothly.

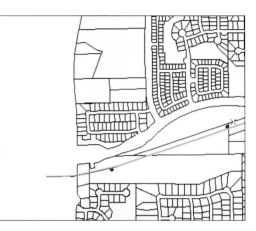

The new rail line is drawn, but the symbology isn't very appealing. Set it as a double-line railroad symbol.

17 Click the symbol for Rail Lines in the table of contents to open the Symbol Selector dialog box. In the Search line, type **railroad multi-track** and press Enter. Several railroad symbols are found. Select the symbol that displays the track with two parallel lines, and then click OK.

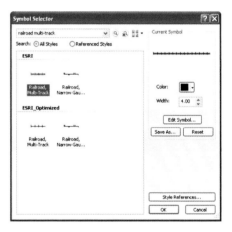

18 Click Clear Selected Features to see the new symbology. This concludes the first part of feature creation. You were able to take the rough schematic of the rail line location and, using the context menu and Sketch Construction tools, draw the line in a fairly precise location.

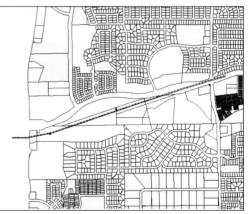

19 Save your map document into the folder where you installed the tutorial data (for instance, **C:\ESRIPress\GIST3\MyAnswers\Tutorial 4–1.mxd.**)

If you are not continuing to the next exercise, exit ArcMap.

Exercise 4–1

The tutorial demonstrated many of the line creation techniques that are available with the regular Sketch Construction tools.

In this exercise, you will repeat the process for the second part of the diagram, which shows the extension of the rail line to the eastern end of the city.

- Start ArcMap and open Tutorial 4–1.mxd.

- Load and zoom to the SecondRailSegment bookmark.

- Snap to the end of the existing rail line and extend it using these measurements:

  - An angle of **12.92** degrees for **375** feet

  - Parallel to the edge of the existing street for a distance of **4200** feet (do not use commas when inserting directions into ArcMap)

- Bonus: Use the Fillet tool on the Advanced Editing toolbar to put a 1500-foot radius on the line.

4–1
4–2
4–3
4–4
4–5

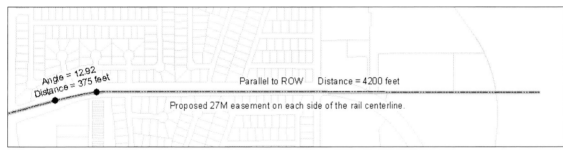

- Save your results as **\GIST3\MyAnswers\Exercise 4–1.mxd**.

WHAT TO TURN IN

If you are working in a classroom setting with an instructor, you may be required to submit the maps you created in tutorial 4–1.

Screen captures of

         Tutorial 4–1

         Exercise 4–1

Tutorial 4–1 review

A successful geodatabase will contain accurate representations of important spatial features (i.e., points, lines, and polygons), important characteristics about these features in the form of attributes contained in an associated data table, and information about behavioral aspects of the relationships among features stored as a topology, as well as built-in rules that enhance the security and integrity of the data. There are several ways to populate these spatial features in the geodatabase. One way is by importing existing features from data contained in shapefiles. As was discussed, the ability to import existing data has tremendous opportunity for cost savings for an organization. However, there will be opportunities where you need to **create new data** as additional needs are identified in the organization. In exercise 4–1, the City of Oleander was interested in building a new light-rail system and had already identified an area within the city for this construction to occur. Existing data of the proposed rail system was not available in a digital format, so it was necessary to create this feature within your geodatabase. By using tools available in ArcMap, you were able to create this digital representation of the rail line from a paper map with exacting specifications.

To ensure that your **data editing** is performed in an organized fashion, it is best to organize how necessary features in the ArcMap session are **displayed** on the screen. In this exercise, you created the proposed rail line feature based on a known start point, and then used existing spatial features contained in other layers to complete the feature edits. As the accuracy of the proposed rail line depended on its relative proximity to other features, it was important that these features were drawn using the appropriate **symbology** and that they were **organized** and **appropriately named** within the **table of contents**, so that important features were recognizable, accessible, and visible to the user. You created **group layers** that combined associated data layers and learned how to customize how and when these features are drawn. You also examined how using **bookmarks** can help define areas of the map that you will need to refer back to often. Lastly, by setting the default **map extent**, you saved valuable time and lessened the potential for getting lost by reducing the map extent from an "outer space" view to the project area. By following the steps in creating an organized map document every time you begin an edit session, it will make your work more efficient and the map document easier to understand.

Before you can edit or create any new data, you must have a container to store these features. The containers that store the data are the feature classes, which may or may not exist within feature datasets within your geodatabase. As you have already examined how to create feature classes and feature datasets in previous tutorials on developing the geodatabase, you should be familiar with the operations necessary to create them. In this tutorial, the **ORTA geodatabase** was provided, including the feature class to contain the proposed rail lines. This line feature class already contained the necessary spatial and attribute information that you needed to place your edits. However, in order for ArcMap to know which feature class to place your edits, you must explicitly select it from the **Create Features template**. Once you have selected the appropriate feature to edit, and the appropriate **construction tool**, you can then begin editing.

ArcMap provides a variety of editing tools on the **Editor**, **Advanced Editing**, and **COGO toolbars**. Shortcuts to these tools are contained in the context menus. While all the tools allow you to create points, lines, and polygons, you will be most successful by matching the task to the most appropriate tool. As illustrated in the exercise to create the proposed rail line, you used several tools in order to create one feature. The ability to successfully use several tools in conjunction with each other depends on your knowledge of what tools are available and how each of these tools can accomplish what you need. The number of tools available for creating and editing spatial features within ArcMap may seem a bit overwhelming at this point, but the more you use these tools in a variety of circumstances, the greater your knowledge regarding the capabilities of each one will become, and the more adept you will become at using them. The main thing to remember is, do not be afraid of testing each tool—that's why there is an Undo button on the toolbar!

STUDY QUESTIONS

1. Please list the different ways to optimize your map document as described in the exercise. Which method is the most beneficial to you and why?

2. What is an edit sketch? How is it different from a permanent feature?

3. List and define five different Sketch Construction tools.

4. Please illustrate an example of creating a sketch employing at least three of the Sketch Construction tools.

4–1
4–2
4–3
4–4
4–5

Other real-world examples

Organizing the way a map document is presented is a critical first step toward efficiently working with the data. A multitude of variables and options exist for managing and portraying the data in a map document. Proper organization of how these are displayed helps keep the focus on the activities necessary to accomplish your objectives. Just as you designed your geodatabase, it is important to organize your map document from the beginning so that you are able to differentiate among features, turn groups of layers on and off together instead of one at a time, and automatically zoom to predefined bookmarks. By doing these things at the outset, you will help reduce clutter within the map document and keep your efforts productive.

Often, you will be working on an editing project where non-GIS users will need to provide input. Say, a planning commission is attempting to establish a political boundary, which up until now has been somewhat ill-defined. In this instance, every boundary that you edit will have long-term implications that the planning commission will have to live with. The ease with which you can move around the map, make your edits, and track your edit progress throughout several meetings depends on how well you organize your map document, and that is determined by the way different layers of information are portrayed.

Additionally, there may be circumstances in the midst of a large data creation project, when your manager will need to assess progress or get additional support from other stakeholders. The way your map document is organized could have a serious impact on those decisions. Illustrating your work efficiently through an organized map document will not only concisely communicate your workflow status, but also go a long way toward garnering support for continued work on the project.

Tutorial 4–2

Using context menu creation tools

The first tutorial in this chapter barely scratched the surface of what can be done to create new features. This tutorial will look at more Sketch Construction tools, as well as methods for using the selection and snapping controls to your advantage.

Learning objectives
- *Set snapping*
- *Set selectable layers*
- *Use the Sketch Construction tools*
- *Create new features*
- *Use the context menu tools*

Introduction

There are many more drawing tools available for creating new features than were demonstrated in the last tutorial. While tutorial 4–1 worked mainly with the context menu, these next techniques will use existing features and more sophisticated measurements to create new features.

One new concept for this tutorial is using the snapping environment. It will give you the ability to use existing features to help draw new features and to set up, with some accuracy, which parts of existing features to use.

You can set each data layer to snap to the endpoints, vertices, or along the edge of a feature. An endpoint is where a feature begins and ends. For a line, this is at each end, with the first endpoint shown as a green box and the last endpoint shown as a red box in the edit sketch. For a polygon, this is the single point where the polygon begins, because a polygon has to start and stop at the same point. This is shown as a red square.

A vertex is a point along a line or the perimeter of a polygon where the line changes direction but does not end. These are shown as green squares in the edit sketch.

Points are easy, because you either snap to the location of the point, or you don't. Setting the snapping of a point to either endpoint or vertex is considered the same thing, and your new feature will snap to the exact x,y coordinate of the point.

An edge is any location along the line or polygon perimeter. This will let you snap anywhere along the length of a line.

In the GIS world, the importance of snapping features cannot be overemphasized. Many of the analysis tools used in GIS analysis require that the features have some connectivity, such as linear referencing or network analysis. If the features are not snapped, no analysis can be done across the line features. A snapping tolerance can be set, which represents the distance the cursor must be from the feature in order for the feature to snap to an existing location. There are also specialized snapping settings for the edit sketch.

Another new concept for this tutorial is setting the selection environment. When a map document contains a large number of layers, making a selection can be troublesome if unwanted features continually get selected. The solution is to selectively set the selection environment. It is also possible to set the selection tolerance, or the distance within which ArcMap searches for features.

With these new settings, it is possible to use existing features to create new ones.

4–1
4–2
4–3
4–4
4–5

Using shortcuts to create features

Scenario   The Oleander Regional Transportation Authority diagram is still not totally entered into the GIS feature datasets. There are some additional linear features relating to property boundaries and parking lots that need to be drawn. You'll refer to their diagrams and use more of the ArcMap Sketch Construction tools to enter these next features. This tutorial also deals with a lot of shortcut keystrokes to invoke various functions in ArcMap. Knowing these can make you very productive in ArcMap and save a lot of time working through the menus. A complete list of shortcuts can be found by searching for "Shortcut Keys for Editing in ArcMap" in ArcMap Desktop Help.

Data   You will be using the same data you copied in tutorial 4–1. The new features to enter will be the right-of-way along the railroad track and the parking lot area. Remember that it's a puzzle with the goal to discover which drawing tools will be able to perform the desired tasks. Note also that in most circumstances, one tool won't draw everything you need. It may be necessary to use a combination of tools and construction lines to draw the final features.

Tools used   ArcMap:

Select Features	Line: Length
Trace tool	Line: Perpendicular
Clear Selected Features	Trim tool
Bookmarks	Copy Parallel tool
Snap to Features	Split Feature tool
Line: Parallel	Fillet tool

The first feature to draw will be the right-of-way for the railroad referenced in the diagram in tutorial 4–1. ORTA will want to purchase 27 meters of right-of-way on either side of the track, and you need to show where that will fall. Since you'll only be dealing with a few of the datasets, you should set the selections to include only the Rail Lines and the Buildings and Paving (Lines) layers.

Set the selections

1 Start ArcMap and open Tutorial 4–2.mxd.

2 At the top of the table of contents, click the List by Selection button. Next to each layer are two buttons. The first will toggle the selectability of the layer, and the second will clear any selected features from the layer. Next to this is a numeric display indicating how many features in the layer are currently selected.

3 Click the buttons to make all the layers not selectable, except for Rail Lines and Buildings and Paving (Lines). Notice that as layers are made not selectable, they are moved into a new pane in the window.

4 Return to the List by Drawing Order button. Now in even the most congested areas of the map, if you click to select a feature, all others will be ignored except these two. As a side note, if you press and hold the Ctrl key and click any selected layer, all the layers will be changed to not selectable. The reverse is also true—pressing and holding the Ctrl key and selecting an unmarked check box will make all the layers selectable. In fact, the Ctrl-click technique works anywhere ArcMap presents a list of check boxes, whether it's selections, layer visibility, activating field names, or any number of other functions. This technique also works with the toggle key in the table of contents to show or hide the legend of layers.

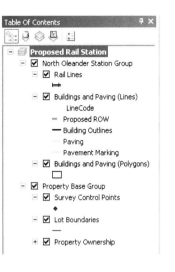

5 Try out the selection by picking the Select Features tool from the Tools toolbar and clicking the railroad track drawn in the last tutorial. Try clicking an area where the rail line crosses other features. Note that only the rail line is selected. After trying the tool in other areas, clear the selected features before continuing.

Start an edit session

According to the diagram, the right-of-way (ROW) exactly parallels the proposed rail line by 27 meters on either side. It would be hard, if not impossible, to modify the information for the rail line to make these ROW lines manually, but with the Trace tool, they can be traced with an offset to provide the desired results.

1 On the Editor toolbar, go to Editor > Start Editing and select the ORTA geodatabase again. The Create Features template will be added to your map document.

Set the edit feature

1 In the Create Features template, click Proposed ROW. In the Construction Tools pane, make sure that Line is selected. Remember to confirm the edit feature selection each time you draw new features. This will keep you from drawing in the wrong layer.

4-1
4-2
4-3
4-4
4-5

Perform the trace

1 Load and zoom to the ORTA Site bookmark. Then on the Editor toolbar, click the Trace Tool button. If it is not displayed, click the drop-down arrow and change the tool to the Trace tool. This tool will exactly trace any feature. It will also jump between features provided that they are snapped, and even duplicate arcs as true curves.

For this instance, you want to trace the line but move it over 27 meters. This is done with the Offset tool and an on-the-fly unit conversion.

2 Press O to open the Offset dialog box. In the Offset distance box, type **27m**. Since "m" is the abbreviation for meters, the distance will be accepted in those units. Click OK. Press O again and notice that the distance has now been converted to feet, which is the map unit for this project. Click OK again.

The tracing will begin at the west endpoint of the rail line and end at the east end. It will be important to snap to the end of the line, so you will need to review the snapping environment.

3 On the Editor toolbar, go to Editor > Snapping > Snapping toolbar. The tool-bar allows you to quickly enable or disable point, endpoint, vertex, and edge snapping. Leave them all enabled for the moment. Since none of the operations you will be doing will require an edge snap, click the Edge button to disable it. Once the snapping is set, the SnapTip in edit mode will indicate which feature type you are snapping to. Snapping can also be disabled, or a special snapping option can be set from the Snapping toolbar.

4 Click the Snapping drop-down arrow and verify that Use Snapping is selected. This enables or disables all snapping. After examining the settings, move the cursor away from the drop-down menu, but leave the Snapping toolbar open. The Snapping drop-down menu also contains three specialty snapping settings: The Intersection Snapping tool will create a snap point where two features cross, even if there is no node at the location or if the features are in different layers. The Midpoint Snapping tool will snap to the midpoint of any line segment. The Tangent Snapping tool allows you to snap a line to be tangent to an existing curve feature. Finally, a fourth Snap To Sketch tool allows snapping to the edit sketch before it is saved as a feature.

In addition to the other snapping tools, the drop-down menu contains a selection to expose more snapping options. First, you'll check to see if the snapping option is working out, and then examine these options to see which one is more advantageous.

5 Move the cursor into the map window and place it over the end of the rail line until you see the SnapTip display. You will notice that the SnapTip is very difficult to see as it displays on top of existing features. There are settings in the snapping options that will help.

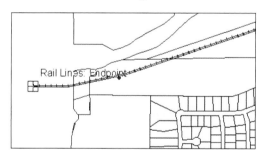

6 On the Snapping toolbar, click the Snapping drop-down arrow and select Options. The options dialog box has two areas. The top pane controls the snapping tolerance and the color of the snapping icon. The toler-ance indicates how close you must move the cursor to a feature before snapping takes place.

4–1
4–2
4–3
4–4
4–5

The lower pane controls the way the SnapTip is displayed. Perhaps changing the SnapTip symbol will make it easier to see.

7 In the SnapTips pane, select the Background check box. This will display a background to the SnapTip to make it stand out against the features.

8 Next, click the Text Symbol button. This will reveal the familiar Symbol Selector dialog box. Click the B button to make the text bold. Click OK, and then OK again to close the Snapping Options dialog box.

9 Move the cursor into the map area and observe the difference in the SnapTip. This one will be much easier to read.

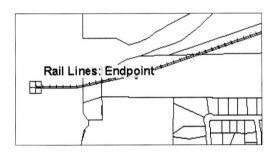

Everything is now set up for the Trace tool to create a new feature offset from the selected feature by 27 meters and to start and stop at the feature's endpoints.

10 Slowly move the cursor to the left end of the selected feature until the SnapTip shows that you are snapping to the endpoint of the rail line. Click that endpoint to start the new feature. Now move the cursor slowly to the right, along the rail line feature. A new edit sketch is being drawn as an exact trace, with a 27-meter offset.

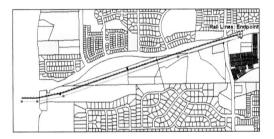

Move to the far end, watch for the SnapTip to display Rail Lines: Endpoint, and click again. An edit sketch will be drawn with a red box indicating the last point entered.

There's another part to this ROW on the north side, but rather than having two different lines, you should make it a single, multipart feature.

11 Right-click to expose the context menu, and then click Finish Part. Notice that the rubber band from the end of the line is gone, but the feature is still in edit sketch mode. This means that you can draw more parts to the feature. The result will be a single record in the table, but there will be multiple graphic representations of that record. This is very useful for features that take multiple parts to represent their entire structure.

12 To create the offset line going back the other way, start where you left off at the Rail Line: Endpoint. Click the endpoint to begin the next part of the feature, and then move along in the opposite direction, again tracing the selected feature. When you reach the left end of the line, double-click. The cursor will not only snap to that end of the line, but also end the feature creation.

13 On the Tools toolbar, click Clear Selected Features to show the symbolized feature.

Next, you'll add another feature to the feature class using some of the special-ized Sketch Construction tools. The trick to using these tools is to know them thor-oughly and determine which ones can be used to achieve your particular goal.

4–1
4–2
4–3
4–4
4–5

You'll put in the perimeter of the parking lot. It is given rough dimensions in the diagram, and shown in relationship to the street and property line, but even then, some of the lines will be sketched without final dimensions.

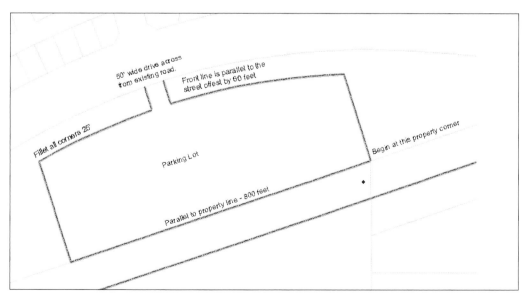

Start with the basic perimeter of the parking lot. The only point that is tied to any existing features is the southeast corner, which coincides with the property corner there. If you snap to that and go parallel to the ROW line, you will be able to draw most of the perimeter.

Along the street, there are no lengths recorded, but it is noted that the front is 60 feet from the street edge and runs parallel to the curve. You can draw that as a separate feature, and then tidy up the corners. The main entry is not located in reference to any other feature, but it is 50 feet wide and appears to line up with a street opposite the main boulevard. You can sketch the entry while keeping it the correct width and tying it to the existing street.

A lot of the context menu tools will be used here, and some of them, such as the Parallel and Perpendicular functions, are available on the Feature Construction toolbar. Be careful with your cursor placement, especially when tracing, and make deliberate movements to help prevent errors.

Zoom to the site

A bookmark has been created for the extent of the new station's parking lot and station building. Add this to your map document and use it to zoom to the site.

1 On the Bookmarks menu, click Manage. Click Load and add the Oleander Station bookmark from the Data folder. Then zoom to it. Close the Bookmark Manager dialog box.

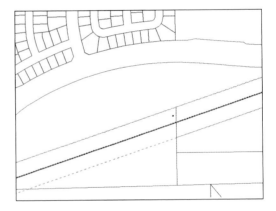

Draw the parking lot

The parking lot will start at the property corner shown in the diagram. The front dimension is given but not the sides. However, since they are shown to stop 60 feet from the edge of the existing road, you'll draw them longer than necessary and clip them later.

1 In this step, you will be snapping to edges. Open the snapping options and re-enable Edge Snapping.

2 In the Create Features template, click Paving and confirm that Construction Tools is set to Line.

3 Select the Straight Segment tool from the Editor toolbar and move the cursor to the property corner to begin. Move the cursor to the property corner shown in the diagram until the SnapTip shows Lot Boundaries: Endpoint.

The edit sketch is started, and the rubber banding is in place. According to the diagram, the pavement runs parallel to the ROW line for 800 feet.

4–1
4–2
4–3
4–4
4–5

4 Select the Constrain Parallel tool from the Feature Construction toolbar, and then find and click the appropriate Lot Boundaries: Edge. Next, right-click and select Length. Enter a distance of **800 feet**. The same procedure could have been done using the Ctrl functions by pressing Ctrl+P and Ctrl+L, and then entering the length.

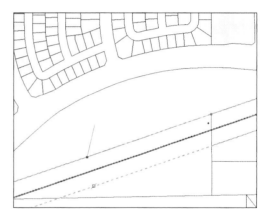

5 Now select the Constrain Perpendicular tool from the Feature Construction toolbar and click the same edge as before. Move the cursor up near the street edge and click to add a vertex. Don't worry too much about placement, since this will be trimmed later. As you recall, the red box over the last vertex indicates that it was the last one entered. And new vertices will be added after this one. What needs to be done, however, is to add a line from the start point that runs perpendicular to the ROW line.

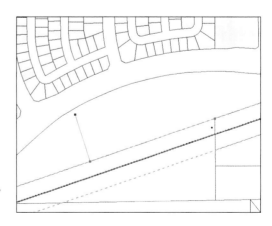

If you reverse the edit sketch, making the first point entered become the last point entered, the new line will continue from there. Confused? Try the command, and maybe the visualization will help.

6 Move the cursor onto the edit sketch line you just drew, right-click, and then click Flip. Now the red square is at the other end of the edit sketch, and new vertices will extend from this point.

Now you can use the perpendicular command again and draw a line out to the street.

7 Use the Perpendicular tool and the same edge as before to constrain the new segment's angle. Move the cursor up near the street edge and double-click to end the sketch. This will also be trimmed later. The north edge of the parking lot runs parallel to and 60 feet offset from the existing street edge.

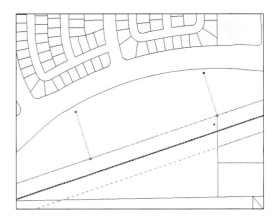

Use the Parallel tool and draw a new feature that goes a little past the one you just drew. Then the overlaps can be cleaned up.

8 Select the Trace tool from the Editor toolbar. Move onto the street ROW until the SnapTip identifies the correct line and begin the trace. Set the offset distance to **60 feet**. If yours starts tracing on the wrong side of the line, change the offset distance to **-60 feet**. When you've completed the trace, double-click to create the feature.

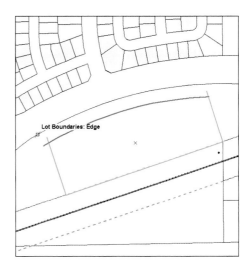

You'll notice that the lines overlap. To clean these up, you will use the Trim tool on the Advanced Editing toolbar. If this toolbar is not present in your session, add it from Customize > Toolbars.

9 With the new curve feature still selected, click the Trim tool on the Advanced Editing toolbar.

10 Move the cursor over the parking lot edge past the intersection on the side you want to trim off and select the line.

Both ends of the line are now trimmed at the selected feature. Next, you want to trim the curve to stop at the end of the pavement perimeter.

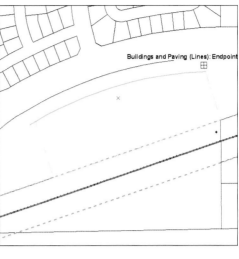

11 Select the pavement edge drawn earlier and use the Trim tool to clip the curve at this boundary.

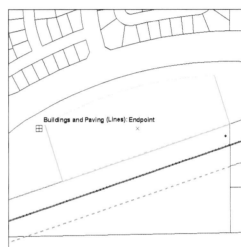

4–1

4–2

4–3

4–4

4–5

Clean up the line work

The perimeter of the parking lot is now drawn. Next, add the entry from the main road. The diagram doesn't show an exact placement, other than it lines up with the drive across the street. It's 50 feet wide, so sketch the first line perpendicular to the parking lot edge, and then copy it 50 feet over.

1 Using the Straight Segment tool, and making sure that the Paving layer is selected in the Create Features template, start a line at the street edge that roughly lines up with the existing drive approach across the street. Pause over the street edge and press Ctrl+E to constrain the sketch to be perpendicular, and then move the cursor into the parking lot so that the sketch overlaps the parking lot edge. Double-click to draw the feature.

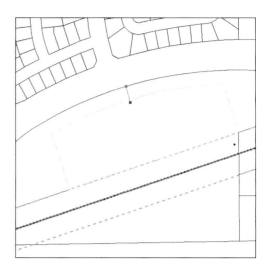

2 Click the Editor toolbar drop-down arrow and select the Copy Parallel tool.

3 Move the Copy Parallel dialog box away from the selected feature. The arrow shows the direction of the line, and looking from the start point to the endpoint will determine left and right. Set the distance to **50** with the side set to Right. Do not worry how the corners are set. Click **OK**. A new line segment is created 50 feet to the side of the selected feature. The options with the Parallel tool will let you control how corners would be made on multisegment lines. You may want them beveled, mitered, or rounded. Examine ArcGIS Desktop Help and experiment with the command to learn more about its function.

YOUR TURN

Go through and clip all the overlapping edges to clean up the drawing. Be careful with the north end of the parking lot. Since it is one single line, if you clip it at the driveway entrance, it will take a major portion away. To solve this problem, use the Split tool from the Editor toolbar to cut it into two pieces first. This is done by selecting the feature, choosing the Split tool, and clicking the line right between the two driveway lines. You won't be able to clip out the existing street, but since this is only a proposal, it will be OK.

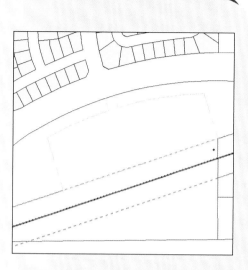

The last thing to be done to clean up the drawing of the parking lot is to round all the corners 25 feet. This is where the Fillet tool on the Advanced Editing toolbar is used. You will select the tool, set the fillet radius, and then select the two features to create the fillet.

4 Select the Fillet tool from the Advanced Editing toolbar.

5 Press R to access the Fillet tool options. Select the check boxes to trim the ends, and use a fixed radius of **25** feet.

6 Select two of the edges to fillet. An edit sketch of the fillet will be shown. Move the cursor around, and you will see that the sketch will jump to one of four places. When it moves to the desired location, click to create the fillet. Then click OK to accept the template Paving.

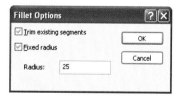

YOUR TURN

Create a 25-foot fillet at each corner of the parking lot. You won't be able to fillet the corners at the existing street, but since this is just a proposal, it would be best not to change the real data. If you have trouble selecting the correct line segments, or segments seem to delete themselves without warning, zoom in to make sure the correct lines are being selected. You may also want to split multisegment lines at their corners before using the Fillet tool.

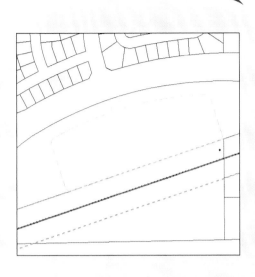

You can see that the context menu and other drawing tools in ArcMap can be a little tricky, and it takes study and practice to determine which ones to use, and where. There are many more tools available, so it is advisable to examine the list of editing commands in ArcMap Desktop Help for more ideas on what the tools are and how to use them.

7 Save your map document as **\GIST3\MyAnswers\Tutorial 4–2.mxd**. If you are not continuing to the next exercise, exit ArcMap.

Exercise 4–2

The tutorial showed more of the editing tools from the standard and Advanced Editing menus.

In this exercise, you will repeat the process for a proposed rail station.

- Start ArcMap and open Tutorial 4–2.mxd.

- Load and zoom to the SecondRailSegment bookmark.

- Add the proposed ROW line for the extension of the rail line.

- Load and zoom to the Commuter Rail Stop bookmark.

- Use the dimensions on the sketch to add the parking lot for the Commuter Rail Stop.

4–1
4–2
4–3
4–4
4–5

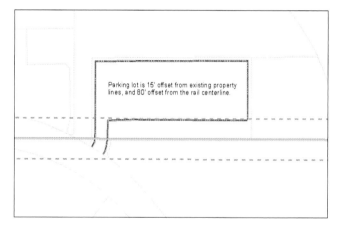

Parking lot is 15' offset from existing property lines, and 80' offset from the rail centerline.

- Save your results as **\GIST3\MyAnswers\Exercise 4–2.mxd**.

WHAT TO TURN IN

If you are working in a classroom setting with an instructor, you may be required to submit the maps you created in tutorial 4–2.

Screen capture of

        Tutorial 4–2

        Exercise 4–2

Tutorial 4–2 review

Creating spatial data in the geodatabase can be a very time-consuming and complicated task. Fortunately, ArcMap provides a multitude of tools that, when learned and applied properly, will help tremendously.

Tutorial 4–2 continued the creation of the Oleander proposed rail station in the geodatabase by illustrating the use of several **standard** and **Advanced Editing tools** to create a right-of-way around the proposed railway and create a parking lot for the terminal station. Before beginning the edit session, you had to do two things that aid the editing process in ArcMap: **set the selectable layers** and **establish the snapping environment**. For a feature to be edited, it must be selected first. Therefore, you need to make sure that the feature you are editing is selectable. The easiest way to do that is to click the **List by Selection button** at the top of the table of contents of your map document and review which features in the list are selected via a marked check box. Not only does this let ArcMap know which features need to be selected, it also prevents any other features from inadvertently being selected for further action.

In the GIS world, the importance of **snapping** features cannot be overemphasized. Many of the analysis tools such as linear referencing and network analysis require that the features maintain **connectivity**. If the features are not snapped, no analysis across the line features can be done. The **snapping environment** provides the "rules of engagement" that will apply throughout the edit session or until otherwise changed for the features you are editing. Components of the lines, polygons, and point features that you create in the geodatabase are referred to as **vertices**, **edges**, and **endpoints**. **Vertices** are much the same as point features, except that vertices are connected by segments and make up line or polygon features. **Point** features and vertices are created using the same methods. **Edges** are any location along the line or polygon perimeter. **Endpoints** are the points at the end of a line segment.

Establishing the snapping environment involves setting a **snapping tolerance**, **snapping properties**, and a **snapping priority**. Although the snapping environment is typically established at the beginning of an edit session, it can be established or modified at any time during editing. In addition, these rules can be intentionally bypassed by using the **right-click context menu** for one-click snapping when editing a feature.

ArcMap can edit only one geodatabase at a time, so a new edit session must identify which geodatabase contains the feature classes that will be edited. When editing several different feature classes in an edit session, make sure to select the intended feature class from the **Create Features template** to ensure that you edit the correct feature class. Failure to do so will place whatever edits you make in the wrong feature class. While editing in ArcMap, take advantage of the **shortcut keys**. There are many time-saving shortcut keys available when editing in ArcMap. In addition, if you need a shortcut that is not available, you can build your own. Look in ArcMap Desktop Help for more detailed information about shortcut keys. Once memorized, the shortcut keys will greatly speed up editing in ArcMap.

This exercise used tools from the **Editor** and **Advanced Editing toolbars**. The Advanced Editing toolbar, as well as any of the other editing toolbars (Annotation, COGO, Cadastral Editing, Dimensioning, Geometric Network Editing, Representation, Route Editing, Spatial Adjustment, Topology, and Versioning), can be added to your map document through the View > Toolbars option. The Editor and Advanced Editing toolbars contain many CAD-like (computer-aided drafting) tools for editing features in ArcMap. The better you know the functionality of each tool, the better you will be able to use several tools in conjunction with each other to complete a task. These tools can be used along with shortcut keys to speed up editing. In this exercise, you used the **Trace tool** in conjunction with the **Offset command** (shortcut key O) to create a line that represented the right-of-way for the proposed rail line that was consistently offset from either side of that line. Being able to use this tool was a very efficient way of accomplishing your task of creating a parallel right-of-way.

Many times, your project may call for specific editing tasks to be used more than others. Task options can be added or removed from the **Feature Construction toolbar** or the Editor toolbar. This can simplify your editing interface. Later in the book, you will learn how to create your own custom toolbars.

4–1
4–2
4–3
4–4
4–5

STUDY QUESTIONS

1. Please list and define five tools from the Advanced Editing toolbar.
2. List and define the three different features set in the snapping environment. Why is it important to specify the snapping behavior during an edit session?
3. Why is it important to define a selectable layer? Explain two different ways of making a layer selectable in ArcMap.
4. Is there another way to create a right-of-way delineation of 27 meters on either side of the proposed rail line? Explain your answer. Would you have arrived at the same result? How?

Other real-world examples

Often, you will need to import a CAD dataset into your geodatabase. Whenever you import CAD data, always look for topological errors. CAD products are often produced with little or no enforced topology. This means that while two endpoints may appear to join on the map, they, in fact, do not, as you will find as you zoom in. ArcMap has many CAD-like feature creation and editing capabilities, and it can enforce topology by use of the snapping environment. If you require a higher level of accuracy in your geodatabase, you will need to use the Sketch Construction tools in ArcMap to edit and fix this data.

Take, for example, a water conservation agency that needed to create a 2,000-foot management area around one of its lakes. It used the Trace tool with a 2,000-foot offset and traced the edge of the lake. While this worked, the agency discovered that for a very complex edge, this might have been accomplished more efficiently by executing the Buffer command in ArcMap to create a 2,000-foot buffer around the lake.

Or take a federal agency that needed to determine the impact of a proposed urban development project near the protected habitat of an endangered bird species. Locations of the proposed development complete with required environmental setbacks were quickly derived from survey plans and digitized into the geodatabase. Sketch Construction tools available in ArcMap were used to develop a range of development scenarios to help minimize impact on the protected species habitat.

The key to mastering editing in ArcMap is to learn when to use which tool. Often, there are several different paths that can be used to arrive at the same goal. In most cases, there is no wrong option—simply a preferred option to help you accomplish your task. Again, this will depend on your experience and skill. As you become more familiar with the tools, your efficiency will increase.

Tutorial 4-3

Exploring different creation tools

The Sketch Construction tools have worked well in the past two tutorials to make new linear features, but how well do they work on polygons? This section will deal with creating polygon features using the context menu functions and a few choices that are specific to polygons.

Learning objectives
- *Use the Sketch Construction tools*
- *Create new features*
- *Use the context menu tools*
- *Work with polygons*

Introduction

All the features used in GIS must be represented by points, lines, or polygons, so it is important that there be good, strong tools for making polygons just as there are for creating lines. Polygons have the unique characteristic that they must start and end at the same place, so locating the start point means that you have also located the endpoint. But for all the other vertices in a polygon, you must rely on the regular ArcMap Sketch Construction tools. And just like the lines, the ability to solve the puzzle of which tools to use when will make you successful at using them.

This tutorial will look at more of the Sketch Construction tools and context menu functions as they are applied to polygons, but the same ideas may be applicable to points and lines as well.

More tools and functions

Scenario   On the next page is the ORTA diagram for the new North Oleander Station, and you can see that the buildings will need to be added. These are represented by polygons, and some information about their locations and dimensions is given. Using the ArcMap Sketch Construction tools and context menu functions, you'll add the rest of the features to the drawing.

The first building to add will be the large one that lies parallel to the tracks. It can then be used as a reference point to add the smaller buildings.

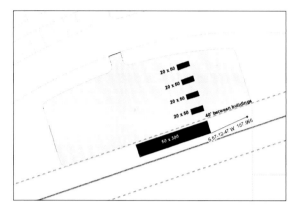

Data   You will be using the same data you copied in tutorial 4–1.

Tools used   ArcMap:

Show SnapTips	Line: Parallel
COGO environment	Line: Square and Finish
Line: Direction/Length	Copy/Paste
Line: Delete Vertex	Selection options
Line: Perpendicular	Trace tool
Line: Length	Symbol Levels

First, you will need to take care of some setup procedures such as setting the snapping environment and the format for the angular input.

Set up editing

There will be a lot of snapping going on in this tutorial, so you should set the snapping environment to include all the features you set in tutorial 4–2.

1 Start ArcMap and open Tutorial 4–3.mxd.

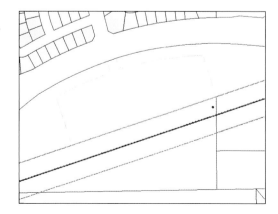

2 Start an edit session and select the ORTA.mdb geodatabase. Next, open the Snapping Environment dialog box and set the snapping options for a background and bold text. Close the Snapping Options dialog box when completed. If you are unfamiliar with how to set these parameters, review the previous tutorial.

Next, you will need to set the units for the angular measurements. The default is polar coordinates, where zero degrees is to the right and angles are measured in degrees counterclockwise. The information provided is in quadrant bearing—a measuring system used by surveyors. The angle describes which quadrant the angle is measured in, with a measurement in degrees–minutes–seconds.

3 In the Editing Options dialog box, click the Units tab. Set Direction Type to Quadrant Bearing and Direction Units to Degrees Minutes Seconds. Confirm that decimal places are set to 0. Click OK. Make sure that Point Snapping on the Snapping toolbar is selected.

4–1

4–2

4–3

4–4

4–5

Draw the buildings

With all the necessary parameters for the COGO environment set up and an edit session started, you are ready to start drawing the buildings. The corner of the building is referenced to the survey control point in survey measurements. You could draw a temporary line to represent the distance, but with some cleverness, it can be done without it.

1 Select the Buildings and Paving (Polygons) layer from the Create Features template. Then select the Straight Segment tool, move the cursor to the survey control point until the SnapTip displays Survey Control Points: Point, and click to start the edit sketch. Invoke the Direction/ Length function by pressing Ctrl+G and enter the dimensions as shown in the diagram at the start of this tutorial: **S 67-12-47 W, 157.966**. Press Enter to store the coordinates and create the vertex in the edit sketch.

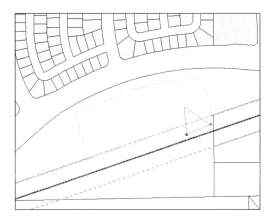

2 Move the cursor back over the start point of the edit sketch, right-click, and then click Delete Vertex. This step has used the survey control point and survey measurement to locate the starting corner of the building but has removed the vertex so that it won't be included in the final feature.

The front and back of the building run parallel to the track, with the sides perpendicular to the tracks and measuring 50 × 300 feet. Using the Feature Construction tools, you should be able to draw the building.

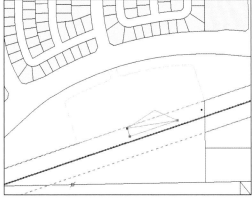

3 Select the Parallel tool from the Feature Construction toolbar, and then select the rail line. Enter Ctrl+L and type a length of **300** feet. Next, click the Perpendicular tool, then the rail line, and enter Ctrl+L with a length of **50** feet. **Note:** You may have to move the cursor a little to get the parallel and perpendicular locks to take effect.

This made half the building. Now it's time for a special feature that can be used with rectangular polygons.

4 Right-click and select Square and Finish from the context menu. This will square off the building and create the feature. It only works with polygons because ArcMap already knows what the last vertex is going to be—it's the same as the start point.

YOUR TURN

The next building aligns with the east end of this building and is offset by 40 feet. It appears to run parallel and perpendicular from this building as well. Use the same technique to draw the new 20 × 50 building.

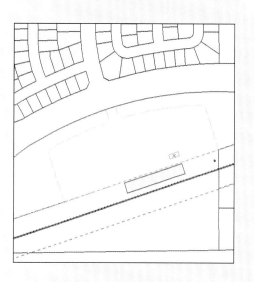

The next buildings are repeats of the 20 × 50 buildings, also set 40 feet apart. Rather than draw them, how about trying a method to copy them and move the copy into the correct position.

5 Make sure the last feature drawn is selected, and then go to Edit > Copy, and Edit > Paste from the main menu. You could also use the icons on the standard toolbar or the Ctrl functions Ctrl+C (copy) and Ctrl+P (paste). Make sure to set the template where the feature will be pasted to Buildings and Paving (Polygons). An exact copy of the building has been added to the map, but it is exactly on top of the first building. You will need to move it, but with some reference to the other building. The crosshair shown in the middle of the selected feature is its attachment point and can be used to snap to other features. The problem is that the crosshair is in the wrong place, though it can be moved.

6 Select the Edit tool from the Editor menu.

7 Move the cursor over the feature's crosshair, and then press and hold the Ctrl key. The crosshair should turn into a little box with arrows around it. Drag the box down until it snaps to the corner of the original station building, and then release Ctrl.

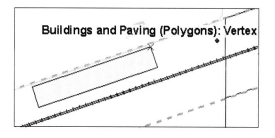

8 Move the cursor back to the selected feature until it becomes a four-way arrow. Then click and drag the feature away from the building you copied. Notice that the crosshair stays in the relative position to the copy of the feature. Move up until the crosshair snaps to the upper-right corner of the second building, and then release the mouse. The building is now copied. This can be a tricky maneuver, but if done correctly, it can be a great time saver. Practice this step until you can do it smoothly.

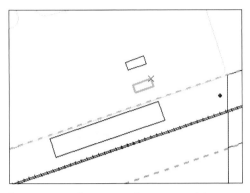

YOUR TURN

Repeat this technique to add two more of the smaller buildings to the north of the two existing ones. They are all 40 feet apart.

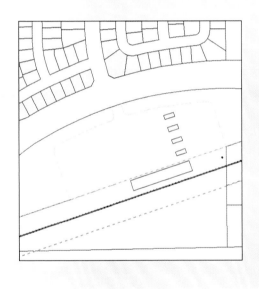

The city planner wants to know how many square feet the parking lot is and wants to symbolize it differently on the map.

Drawing the feature into the Buildings and Paving-Polygons feature class will be an easy trace task. But there needs to be a unique way to identify the parking lot so that it can be symbolized differently. There is a field called Description in the table. The first building is the station, and the out buildings are bus shelters. If these are added, and the parking lot is given a unique description, it can be symbolized that way.

9 Right-click the layer name Buildings and Paving-Polygons in the table of contents, point to Selection, and click Select All. Next, click the Attributes button on the Editor toolbar to open the Attributes dialog box. There are five features shown. One is the station, and the others are bus shelters. A fast way to populate the attributes would be to populate them globally to Bus Shelter, and then discover which one is the station and change it.

10 Click Buildings and Paving-Polygons at the top of the dialog box. When this is selected, any changes you make will be applied to all selected features. Next, click in the lower pane in the empty line next to Description. Type **Bus Shelter** and press Enter.

11 Now in the upper pane, click each of the features and watch in the drawing until the station flashes. When you find it, change its description to Station. Close the Attributes dialog box.

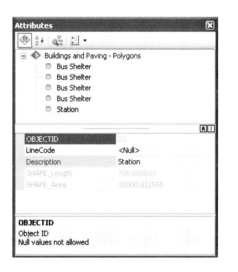

With that taken care of, you can draw the polygon for the parking lot. But you want to make sure not to forget to set its attribute value. You can set an option that will automatically populate the attribute information each time a feature is drawn.

4–1
4–2
4–3
4–4
4–5

12 Right-click the Buildings and Paving-Polygons layer in the Create Features template and select Properties. On the empty line next to Description in the Attributes pane, type **Parking Lot**. Click OK. This value will now be the default value for all new features entered. You're only drawing one, but if you were drawing many more parking lots, this would be a big time saver. Always remember, however, to change the default value before drawing other feature types.

13 Select the Trace tool, make sure the Buildings and Paving-Polygons layer is selected in the Create Features template, and perform the trace.

The feature was drawn correctly, but now it obscures the bus shelters. To fix this, you will change the symbology for the feature.

14 Open the properties of the Buildings and Paving-Polygons layer and click the Symbology tab. Set the symbol classification to Unique values with Description selected as the Value Field. Add all the values, clear the <all other values> check box, and select appropriate colors.

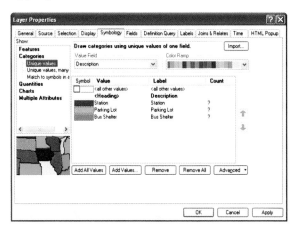

15 Click Apply and examine the results in the map.

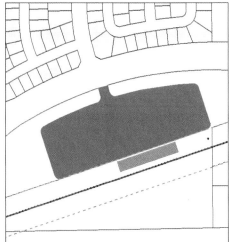

Well, that almost worked. The parking lot did, in fact, draw in the chosen color, but now it obscures the bus shelters. There is a way to fix this by using an advanced setting for symbology.

16 In the lower right of the Symbology Editor, click the Advanced drop-down arrow, and then click Symbol Levels.

17 Select the check box to activate the symbol levels, and move Parking Lot to the bottom of the list. To learn more about Symbol Levels, click About Symbol Levels.

18 Click OK to close the dialog box, and then OK again to close the Symbology Editor. **You may have to refresh the drawing to see the results.** This seems to have done the trick. The parking lot is drawing at the bottom of the layers, and the bus shelters are now visible. Symbol Levels is a very useful feature to control the layering of features in a single feature class. Previously, the features would be drawn in order of their ObjectID number, but now that control has been turned over to the user.

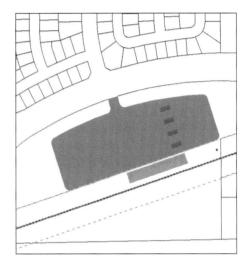

4–1
4–2
4–3
4–4
4–5

This concludes the work with the ORTA diagrams. This exhibit can now be shown to the City Council for reference and used to select the number of households within a specified distance of the station or to examine the utilities in the area to determine if there is any work the City may have to complete.

For more tips and tricks in ArcMap, search ArcGIS Desktop Help for Tips and Tricks within ArcGIS. There is a lot of information there, along with links to white papers, other tutorials, and a printable PDF guide to the keyboard shortcuts.

19 Save your map document as **\GIST3\MyAnswers\Tutorial 4–3.mxd**. If you are not continuing to the next exercise, exit ArcMap.

Exercise 4–3

The tutorial showed how to use the same context menu tools to draw polygons as you would for points and lines. It also demonstrated the Symbol Levels options.

In this exercise, you will repeat the process for the buildings and parking lot at the commuter station.

- Start ArcMap and open Tutorial 4–3.mxd.

- Load the bookmark Commuter Rail Stop, if it is not already loaded, and zoom to it.

- Use the trace technique to draw the parking lot.

- Draw the buildings as shown.

- Populate the attribute table with the correct descriptions.

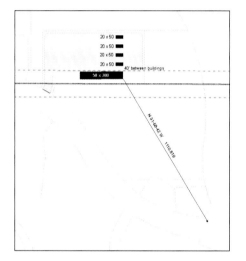

- Save your results as **\GIST3\MyAnswers\Exercise 4–3.mxd**.

WHAT TO TURN IN

If you are working in a classroom setting with an instructor, you may be required to submit the maps you created in tutorial 4–3.

Screen captures of

Tutorial 4–3

Exercise 4–3

Tutorial 4–3 review

Creating **polygons** in ArcMap is very similar to creating lines. With polygons, the difference is that the feature you create will close at the same location that it began. All the feature editing thus far has relied on the same initial steps:

- Adding the feature class to your map document
- Selecting the feature class for editing
- Establishing your snapping environment
- Specifying the feature to draw and tool to use in the Create Features template
- Beginning the edit session

As you could see, the same **tools**, **context menus**, and **shortcut keys** available when creating line features are available when **creating polygons**. Creating polygons was a good opportunity to continue to practice and memorize these tools for later use.

4–1
4–2
4–3
4–4
4–5

You are continuing to learn many new tips and tricks that will make your editing job easier and more accurate. Some of these appear as **SnapTips** as you edit or create a feature and encounter another feature. Something as simple as showing what feature you are snapping to helps reduce confusion and prevent error, particularly when you are showing several different layers in your map document or have several features in close proximity to each other. As you continue to work with ArcMap, you will uncover many of these tricks and begin to implement them routinely. A good source can be found by searching ArcMap Desktop Help for "Tips and Tricks within ArcGIS." A few minutes researching these tips and tricks will save you hours in the long run.

Because you were relying on a survey document to create your station and associated buildings, you needed to set the **angular measurements** for your sketch polygon consistent with the measurements contained in the survey. This was necessary so that you could accurately place the building in the proper location, as well as construct the shape of the station as intended in the survey plans. Since you used the bus station as a relative location to place the associated bus shelters, it was critical that the location and dimensions of the bus station be as accurate as possible.

Having the ability to edit attributes of features during an edit session is a very convenient way of **attributing data** as you create it. By selecting all the shelters, you were able to attribute multiple buildings simultaneously. This not only prevented potential errors in editing the data, but it also prevented you from having to go in afterward and add the attribution.

A little common sense goes a long way when you are creating features in ArcMap. Think through your objectives and develop an **editing plan** ahead of time to outline your objectives and examine what tools are available in ArcMap. As you become more familiar with how all the tools work in ArcMap, you will be able to combine this knowledge with some ingenuity to figure out how to solve a puzzle in the most efficient manner.

STUDY QUESTIONS

1. In this exercise, the distance between buildings to be created was given. You were able to use this information to determine the length and width of the buildings. How could you use the editing tools to create these buildings if you were only given the coordinates for the corners of each building?

2. Could you have used lines instead of polygons to represent the buildings? What would be an advantage and a disadvantage of using this approach?

Other real-world examples

The procedure for transferring features from a surveyed drawing into the geodatabase is a popular means of getting data into the geodatabase. A local redevelopment company can use existing survey data to map out the locations of buildings on one or several pieces of property slated for development. Once within the geodatabase, you can analyze the locations of these buildings with respect to existing nearby environmental liabilities to get an idea of potential impacts on the property.

A pipeline company can digitize the locations of existing utilities from survey documents on or around the pipeline right-of-way corridor to help determine potential damage to these facilities during pipeline replacement activities or alleviate damages to the pipeline during utility replacement activities.

The U.S. Army Corps of Engineers can evaluate the impact of a new development within a certain floodplain. Surveyors have drawn plans illustrating the proposed development and submitted the plans to the agency. The agency can transfer this data into the geodatabase using the Sketch Construction tools so that flooding scenarios can be developed to ensure that effective upstream or downstream mitigative procedures will be required as a result of this development, and total costs for the project can be estimated.

As you have learned, there are many editing tools in ArcMap to facilitate data input. Once data has been entered in ArcMap, it becomes available for the powerful spatial analysis tools of ArcGIS. The initial work of accurately digitizing this data into the geodatabase pays for itself when robust spatial analysis techniques are performed in an automated GIS.

Tutorial 4–4

Using coordinate geometry

One of the best and most accurate sources of data for GIS is coordinate geometry (COGO) data, or survey data. This data is typically very precise and contains informa-tion to lock down its location on the ground in real-world coordinates. Transitioning survey data into your GIS can help improve the accuracy of the existing data as well.

4–1
4–2
4–3
4–4
4–5

Learning objectives

- *Set up the COGO environment*
- *Understand survey data*
- *Create new features from COGO data*

Introduction

When property is surveyed, two things are important to record: The first is its location in real-world coordinates. Adjacent surveys and control monuments are used to reference the corners of the property so that it can be found again. The second is the property boundary. A good boundary survey will describe exactly how much property is contained.

Because the survey industry is tightly regulated, with surveyors required to have a license, the description they provide of property is held up as the legal record. Any property bound-ary disputes can be resolved by going back to the property's survey, or legal record, and determining the property's location and boundary.

In GIS, it is important to store this information as correctly as possible, and to store the legal record of the property as shown in the survey. But when adding survey data to GIS, this is not always possible. Steep grades and older, less-accurate surveys may play a part in making the assimilation of survey data into a seamless GIS dataset impossible. Surveys from the 1800s may still survive as the legal record for some property in Texas, for instance, but matching that to the survey technology that exists today may cause discrep-ancies. So, it is common in the GIS world to stretch, or rubbersheet, data to fit into what we can see is the correct area, even if it disrupts the information in the legal record.

Fortunately, there is a way to store the legal record yet still make a seamless GIS dataset. In the ArcMap COGO tools, there is a function to put additional fields in your feature class to store the legal record. Then if rubbersheeting has to be done to make the survey fit the map, the legal length and angle will be preserved.

In this tutorial, you'll set up a geodatabase to accept survey data and preserve the legal record. Then you'll enter COGO data and see what it takes to make it fit into the existing map.

Using the COGO tools

Scenario   A new subdivision called the Landing at Eden Lake has been platted in the City of Oleander. The drawings of record, including the legal survey, have been submitted so that you can include this property in the parcel base for Oleander. The process will be to update the existing parcel lines feature class to include the COGO fields to preserve the legal record, and then draw the perimeter of the new subdivision. Later, you'll georeference the scan of the plat to this perimeter and trace the rest of the property lines.

Data   You'll use the LotBoundaries feature class from the City of Oleander geodatabase in the Data folder. Also provided are the following scans of the engineer drawings and plat:

The Landing:
Cover sheet
Plat
Sewer Line A
Sewer Line B
Sewer Line C
Site plan
Storm Drain A
Storm Drain B
Storm Drain C
Storm Drain Laterals A
Utility layout
Water plan

You'll refer to these for the information required to add the surveyed plat to the GIS data.

Tools used   ArcCatalog:
Add COGO Fields
New Feature Class

ArcMap:
COGO environment
Traverse tool

This COGO work will require that you add the COGO fields to the existing LotBoundaries feature class. Then even if some adjustments are made, which may alter the length and angle of the actual geometric feature, the legal descriptions will be preserved.

Add the COGO fields

1 Start ArcCatalog and browse to the MyAnswers folder. If you have not completed the previous tutorials that created the LandRecords geodatabase, copy the City of Oleander geodatabase from the Data folder to your MyAnswers folder. Select the LotBoundaries feature class in the PropertyData feature data-set and click Properties. Click the Fields tab and investigate the attributes. Close Properties. There are only a few fields there, and all but LineCode were added automatically by ArcCatalog when you created the feature class.

4–1
4–2
4–3
4–4
4–5

There's a command that will add the required COGO fields to this feature class, but it doesn't appear on the regular ArcCatalog menu. You'll add it, and then run it to add the necessary fields.

2 On the main menu, go to Customize > Customize Mode. Click the Commands tab and scroll down to Geodatabase tools. In the right pane, look for the Create COGO Fields tool.

3 Select the Create COGO Fields tool and drag it up to the toolbar near the Identify button, and it will drop into place. A black vertical bar will show where the tool will be placed when you release it. Click Close.

4 Make sure that LotBoundaries is still selected, and then click the new Create Cogo Fields button. Click OK to acknowledge that the COGO fields were added successfully, and then click Properties and click the Fields tab again. After noticing the changes, close Properties. All these additional fields will be used to store the legal information entered from the plat. Why store all this data? Because what if you wanted to show the legal survey angle and distance for each property line? But when you added the survey, you had to tweak it 0.3 feet to make it fit the existing data. While this may not seem like a lot, by doing this small change, you will have changed the angle of the line, and the SHAPE_Length value. If you used this value or angle as a label, it would not reflect the legal record. The COGO fields, however, are populated by the COGO tools, and their values will not change if the geometry is altered.

Eventually, this data will need to be represented as polygons, but the polygons won't be able to preserve the survey data. Their perimeter is one feature and can only hold one angular direction and one measured distance, or in survey terms, one set of bearings and distances. By making them lines first, you can store the survey data, and then later use the lines to create the polygons.

You will want to start entering the survey data, but it would be quite dangerous to do so in the full LotBoundaries feature class. This might take some time to complete, and others might be using this feature class for other projects that might be hampered while you work through the process. Instead, you will make a new feature class with the same field structure, do your work there, and add it to the LotBoundaries feature class when it is complete.

5 Right-click the PropertyData feature dataset, point to New, and click Feature Class. Give it these parameters:

- Name: LandingPerimeter
- Alias: The Landing at Eden Lake
- Feature Type: Line
- Attributes: Import from LotBoundaries

Note: If you are working with a copy of the City of Oleander geodatabase rather than the geodatabase you created in the earlier tutorials, name your feature class TheLandingBoundary to avoid duplicating an existing file.

6 Close ArcCatalog. With the new feature class created to store COGO data, it is time to set up ArcMap to record the COGO data.

Set up the COGO environment

1 Start ArcMap and open Tutorial 4–4.mxd. Add the new feature class you just created to the table of contents.

The next steps are necessary to make sure that the ArcMap COGO tools will perform as expected and store all the survey data in the correct fields.

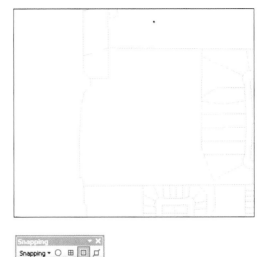

2 With the Editor toolbar active, start editing, making sure to select the geodatabase containing The Landing at Eden Lake (or LandingPerimeter) feature class. Then go to the Snapping toolbar and set the only snapping type allowed to Vertex Snapping. Close the Snapping toolbar.

3 Again from the Editor toolbar, open the Editing Options dialog box. Click the Units tab and set Direction Type to Quadrant Bearing and Direction Units to Degrees Minutes Seconds. Verify that the number of decimal places is 0. Click OK.

4 On the main menu, click the Customize tab, point to Toolbars, and click COGO to get the COGO tools on screen.

5 Finally, click The Landing at Eden Lake in the Create Features template. This completes the setup for the COGO tools.

Use the COGO toolbar

Next, take a look at the plat to see what data will be entered.

1 Open the image file The Landing—Plat from the Data\Images folder. From the enlargement of the lower right as shown, you can see that the perimeter of the subdivision is described in a series of angle–length pairs. These are called bearing and distance, with the bearing measured in the quadrant bearing degrees, minutes, and seconds that you set in the Editing Options dialog box. These measurements are in bold print around the subdivision's boundary line.

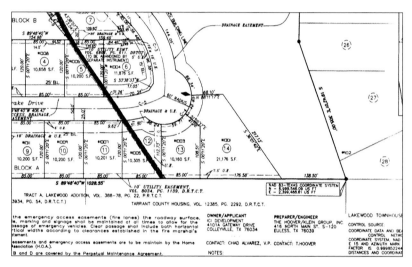

2 You'll also notice in the same image a box referencing a set of ground coordinates given in the Texas State Plane, NAD 83 coordinate system, which you may enlarge to see more clearly. Since this is the same coordinate system as the feature class, adding these numbers as an absolute x,y coordinate will place the corner in its exact location, thanks to an alert surveyor.

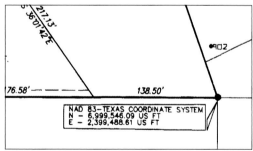

3 Lastly, over in the lower right of the plat, there is a comment about a ground-to-grid conversion for this plat. Surveyors measure on the ground, and the map is in the Texas State Plane coordinate system (a grid), so there is a conversion to go from one to the other. It is shown as using the combined scale factor and the convergence angle.

4 Go to Editor > Options again, click the Units tab, and add the Ground to Grid Correction conversion factor detailed in this note. This will perform the necessary angular and scale conversions on the fly as the data is entered. Click OK once the parameters are entered. Not all surveys will contain the grid coordinates for the corners or the ground-to-grid conversion. The City of Oleander has written into its ordinances that this information must be provided on every plat. Good surveyors will have no problem providing this detail since their CAD systems can easily calculate the values.

To enter the subdivision perimeter using the survey data, you will want to start at one of the absolute x,y coordinate pairs and proceed clockwise around the plat. This is the standard practice in the survey industry, and it will ensure that all the bearings are measured in the correct quadrant.

5 On the COGO menu, select the Traverse Tool button to open the Traverse dialog box. Click Template and select The Landing at Eden Lake. At the lower right, select the Closed Loop check box, which means that the set of data you enter will return to the start point and complete the perimeter of the survey. Next, click Edit and enter the x,y coordinates of the start point that you observed on the plat earlier. Click OK.

The data entry process has started. Now you will follow the image around the perimeter of the survey and enter the bearing and distance necessary to create each side. Remember to go clockwise around the survey. The bearings are entered as the letter N or S and a space, then the degrees-minutes-seconds angle separated by dashes, then a space, and finally a trailing E or W. The spaces aren't necessary but will make the data easier to follow. You may also capitalize or lowercase the letters. The distances should be entered using numbers only, without commas.

6 At the upper-left corner of the Traverse dialog box, verify that the input method drop-down list is set to Direction–Distance. Enter the first set of data and click Add.

- S 89-48-40 W, 1028.55

7 Continue around the survey and enter the following sets of data, clicking Add after each one:

- N 8-1-39 W, 67.23
- S 89-57-21 W, 18.00
- N 0-12-39 W, 231.89

All the entries so far have been lines, but the next segment of the survey boundary is a curve. The ArcMap Traverse tool can build and store a true curve based on the geometric data provided in the survey. The survey will typically include a radius, arc length, sweep angle, chord length, and chord bearing. The accompanying diagram shows what each of these terms mean.

To create the curve, only a certain number of these elements need to be entered, depending on whether or not this is a tangent arc. Normally, the surveyor will identify tangent arcs in the written description of the boundary, but since this was not included, you should use the Nontangent Arc tool to make sure it is entered correctly.

The information on the plat for the arc is identified by abbreviations such as

> Δ = Delta or sweep angle
>
> R = Radius
>
> L = Arc length
>
> Ch. Brg = Chord bearing, followed by the chord length

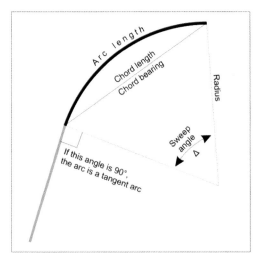

There are no standard abbreviations for these terms, so you may have some guessing to do from plat to plat. Examine the values, and you should be able to determine what each value might represent. For instance, anything listed as a quadrant bearing must be the chord bearing. If two lengths are given, the shorter is the chord length, and the longer is the arc length. Anything shown as an angle in degrees-minutes-seconds but with no letter prefixes or suffixes must be the delta or sweep angle. Not every curve will have all this information given, but you'll see in the next step that not all these values are necessary. You can customize the curve dialog box to match the data that is present.

Once you've determined all the values, you can continue entering the data from the plat.

8 In the Traverse dialog box, use the drop-down arrow at the upper left to change the entry type to Curve. Notice that one of the choices was Tangent Curve, which you would use if the spring point of the curve were tangent to the last line segment drawn. There are several boxes for data entry shown. If you click a few of the drop-down arrows, you'll see that the boxes can be changed to accept a variety of information.

4–1

4–2

4–3

4–4

4–5

9 Change the first drop-down choice to Arc and enter the arc length of **230.08** in the box to the right. Then enter the radius of **2805.13** and chord direction of **N 2-08-21 E** in their respective boxes. Finally, you must decide if the curve goes to the right or left. Imagine standing at the starting point of the curve and looking down its length. Is the center point on your right or left? In this case, it would be on the right, so change the last parameter to Right and click Add.

10 Change the entry type back to Direction-Distance and continue around the survey, adding the following data to create the perimeter:

- N 4-29-21 E, 158.22

- N 44-57-21 E, 95.61

- N 0-2-39 W, 80.00 (**Note:** This was entered incorrectly so that the editing tools can be demonstrated later)

- N 34-15-26 W, 12.42

- N 89-49-25 E, 147.68

- N 0-10-35 W, 227.72

- S 89-15-30 E, 688.07

- (**Note:** A direction-distance pair was intentionally missed here so that the editing tools can be demonstrated later)

- S 0-54-0 W, 731.98

- S 19-42-1 E, 305

Correct the COGO errors

It's apparent from looking at the finished line that the start points and endpoints are not coincident, so there must be a mistake in the entry somewhere. From going back over the data, and knowing about the intentional error that was entered, you can see that the distance for entry number 8 should be 40 instead of 80. This can be corrected without having to reenter all the data.

1 In the Traverse dialog box, highlight line 8 where the error was discovered. Then click the Edit an Entry button (hand and paper icon) and correct the distance in the resulting Course dialog box, making it **40**. Click OK and review the line in ArcMap again.

The visual analysis of the results suggests that there might still be an error, but clicking Adjust at the lower right of the dialog box would also reveal there is an error. If you try this, be sure to click Cancel before going back to the Traverse tool to correct the problems.

2 It was determined that one of the direction-distance pairs was missed. It was **S 89-4-0 E, 48.98**. Enter this data and click Add.

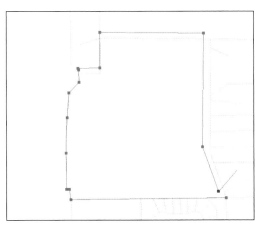

3 The new data was added to the bottom of the list as element number 15. It should be the 13th entry. Click it and use the Up arrow at the right of the dialog box to move the entry up to number 13. Notice that the edit sketch in the map now looks OK.

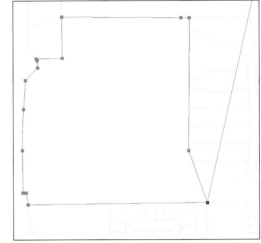

4 To examine the results, click Adjust. This will look at the difference between the start point and the endpoint to determine whether the survey closes correctly. A little error is expected, but the values should be within one-hundredth of a foot. This sketch looks very good.

5 Click Accept to lock in the results and create the new features, and then close the Traverse tool.

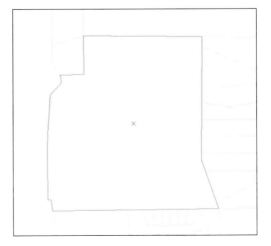

Verify the COGO fields

1 Open the Attribute table of The Landing at Eden Lake layer. Notice that all the COGO fields have been populated, including the data for the curve. Also, compare the COGO Distance and SHAPE_Length fields. Close the Attribute table. This process has preserved the legal record in the Attribute table. The difference you saw between the SHAPE_Length and COGO Distance fields is because the Adjust tool made a slight change to force the survey to close exactly. If you were to label the lines with their length, you would want to use COGO Distance so that the true legal length is shown. While using the COGO tool, ArcMap created a new line feature for every direction-distance pair entered. This enables you to store the survey data for each individual line.

Table							
The Landing at Eden Lake							
COGO Direction	**COGO Distance**	**COGO Delta**	**COGO Radius**	**COGO Tangent**	**COGO ArcLength**	**COGO Si**	
S 89-48-40 W	1028.55	<Null>	<Null>	<Null>	<Null>	<Null>	
N 8-1-39 W	67.23	<Null>	<Null>	<Null>	<Null>	<Null>	
S 89-57-21 W	18	<Null>	<Null>	<Null>	<Null>	<Null>	
N 0-12-39 W	231.89	<Null>	<Null>	<Null>	<Null>	<Null>	
N 2-8-21 E	230.016	4-41-58	2805.13	115.105	230.08	R	
N 4-29-21 E	158.22	<Null>	<Null>	<Null>	<Null>	<Null>	
N 44-57-21 E	95.61	<Null>	<Null>	<Null>	<Null>	<Null>	
N 0-2-39 W	40	<Null>	<Null>	<Null>	<Null>	<Null>	
N 34-15-26 W	12.42	<Null>	<Null>	<Null>	<Null>	<Null>	
N 89-49-25 E	147.68	<Null>	<Null>	<Null>	<Null>	<Null>	
N 0-10-35 W	227.72	<Null>	<Null>	<Null>	<Null>	<Null>	
S 89-15-30 E	686.07	<Null>	<Null>	<Null>	<Null>	<Null>	
S 89-4-0 E	48.98	<Null>	<Null>	<Null>	<Null>	<Null>	
S 0-54-0 W	731.98	<Null>	<Null>	<Null>	<Null>	<Null>	
S 19-42-1 E	305	<Null>	<Null>	<Null>	<Null>	<Null>	

I◀ ◀ 1 ▶ ▶I (15 out of 15 Selected)

The Landing at Eden Lake

There seems to be a little error along the western side, but this is probably because of some new highway work that required additional right-of-way clearance. The new lines fit very well, although some editing may be done later to get a better fit. This may further change the value of SHAPE_Length, but it will never change the true COGO_Distance field.

2 Save your map document as **\GIST3\MyAnswers\Tutorial 4–4.mxd**. If you are not continuing to the next exercise, exit ArcMap.

As you look at more surveys and are asked to deal with the curves, you come to realize that they don't always include all the data you need. One of the ArcMap COGO tools can help. It's the Curve Calculator located on the COGO toolbar. You can enter any two known pieces of information about a curve, and the Curve Calculator will compute the rest. Try out the curve data from the tutorial and see how close the calculator gets.

Exercise 4–4

The tutorial showed how to set up and perform COGO data entry.

In this exercise, you will repeat the process with another subdivision in Oleander.

- Start ArcMap and open Tutorial 4–4.mxd.

- Load and zoom to the Running Bear Estates bookmark.

- Use the COGO tools to enter the perimeter of the subdivision. A scan of the plat is available in the Data\Images folder. At the top is a list of coordinate pairs for the extreme corners of the plat. Use any one of these to start the traverse, and then continue entering directions and distances clockwise from that point.

- Repeat the COGO entry with the following plats:

    - Bear Creek Estates IV

    - Creek Wood Estates

    - Little Bear II

Bookmarks and scans of these plats are available with the provided data.

- Save your results as **\GIST3\MyAnswers\Exercise 4–4.mxd**.

4–1
4–2
4–3
4–4
4–5

WHAT TO TURN IN

If you are working in a classroom setting with an instructor, you may be required to submit the maps you created in tutorial 4–4.

Screen captures of

> Tutorial 4–4
>
> Exercise 4–4

Tutorial 4–4 review

When property is surveyed, two things are important to record: its location in real-world coordinates and the description of the property boundary. A legal survey is the legal record of the property, and it is important to store the details of the survey (the legal record) in the geodatabase.

COGO tools available in ArcMap allow you to transcribe the survey measurements into the geodatabase directly from the survey and duplicate the line work from the survey notations. Often, such measurements do not line up with the existing GIS land base, and the survey boundaries must be adjusted to fit the GIS land base. This process is called **rubbersheeting**. When you rubbersheet the surveyed boundaries to fit the land base, the positions of these boundaries change and no longer reflect the original data submitted in the survey. In such cases, it is important to use the COGO tools in ArcMap to provide the fields within the database to **preserve these original measures regardless of positional adjustments to the GIS land base**. You did this by having the COGO tools insert the appropriate fields to contain this information directly within the feature class, thus **preserving the legal record of these boundaries**.

The **COGO Traverse tool** allowed you to input coordinate pairs to simulate the original survey data. There were, however, a few items that you needed to confirm that involved how the data was originally surveyed. Most surveyors survey boundaries by using "ground"coordinates. This is done to preserve the highest level of accuracy possible in the survey. It was not until the GIS folks attemped to input the data directly into a projected map that it appeared the data was not positioned correctly. This positional inconsistency with the projected data was because the data surveyed in ground coordinates does not take into consideration a projection until a projection is applied to the data. You can remember from your introductory GIS courses that when you apply a projection to data, you attempt to flatten the data onto a plane. Regardless of how you attempt this through the use of a projection, positional adjustments to the data occur. To account for this, most surveyors will have their CAD programs calculate the difference between the ground and the proposed projection, or grid. Most of the time, given a small geographic extent, the **ground-to-grid conversion factors** are very small. Nonetheless, even the smallest conversion factor will have an impact and can be accounted for using the COGO tools in ArcMap.

STUDY QUESTIONS

1. When you began digitizing The Landing at Eden Lake feature class, you set up the snapping environment to snap to vertices. Why wouldn't you want to snap to edges or endpoints as well?

2. How can you determine what angular units were used in the source survey?

3. Why is it important to consider whether the survey was submitted in grid or ground coordinates? How would you account for this difference in the feature class?

4. What happens if you make a mistake typing the dimensions of a coordinate pair? How can you fix this while editing?

4–1
4–2
4–3
4–4
4–5

Other real-world examples

The editing tools available in ArcMap resemble tools contained in CAD packages. Although they may function similarly, products derived from CAD and GIS systems can vary greatly. The most important aspect of a feature in CAD is the graphic line work and annotation representing and describing it. While the graphic aspect of the feature is certainly important in the GIS world, the capabilities of the GIS to store attribute information about the feature in a related database are critical for consequent spatial analyses. COGO tools in ArcMap provide many of the same precise drafting capabilities found in CAD systems yet will typically result in a GIS-compatible product.

Let's say a municipality needs to illustrate the temporal change of original property boundaries for an existing landfill. Survey calls are available from the original paper survey that described the boundary as it existed during the original and subsequent surveys. Using COGO tools in ArcMap, the boundary can be accurately transcribed from each of the paper surveys into ArcMap. The resulting GIS layers from each survey are overlaid for comparison and further GIS analysis.

A local planning agency needs to derive a floodplain/landownership easement from multiple surveys conducted fifty years ago. The objective of the project is to identify the most appropriate location to build a jogging trail within the flood easement of a lake but outside the 100-year Federal Emergency Management Agency (FEMA) floodplain delineation. COGO tools in ArcMap are used to transcribe the landownership survey's original flood easement boundaries. A FEMA floodplain map of the area is acquired, and ArcMap is used to overlay the 100-year floodplain with the transcribed flood easement data to identify the most appropriate location for the proposed jogging trail.

Tutorial 4–5

Georeferencing and tracing

Using the COGO tools to input data is the most precise method you can use. Accurate survey data is transferred to the GIS feature class in a manner that preserves that accuracy. However, it is rather time-consuming. It may also be the case that the plat doesn't contain enough data to use the COGO tools for every line. Older plats don't always have the survey descriptions of internal lot lines. It is possible in such cases to scan the image, put the scan into the GIS map document, and trace the lines that may not have clear dimensions.

Learning objectives
- *Rectify images*
- *Use the drawing tools*
- *Store COGO data*

Introduction

A scanned image of a plat, or even a PDF document from the surveyor, will almost never have a spatial reference. If you were to add this image to your map document, it would appear in some unknown area with no relationship to your data. The solution is to create a spatial reference for the otherwise unusable scan.

To do this, use the Georeferencing toolbar in ArcMap. This toolbar contains the tools necessary to locate the image in real space and lock it into its real-world coordinates. You will need to first determine some common points between the scan and your data. Then when you activate the tools, you will create control points from a point on the scanned image to that point's real location in the GIS dataset. You will need at least two points to determine the proper location, scale, and rotation, but the more points you can ascertain, the more accurate the placement will be.

For plats, you can add as much data in COGO coordinates as possible, and then use the georeferencing tools to locate the plat in real space and trace the rest of the lines. While this is not as accurate as a fully coordinated plat, it's sometimes the only way to get older data transferred.

Filling in the remaining data

Scenario   The Landing at Eden Lake plat had its perimeter drawn with the COGO tools. While that particular plat could be entered entirely with COGO dimensions, for the sake of this tutorial, it will be said that the remainder must be traced in. There is an image of the plat that contains only the area with property lines, excluding most of the notes. This will be georeferenced to the map document, and then the remaining property lines will be traced.

It is important to note that even though the new features will be traced from the image, it will still be prudent to use all the snapping and other tricks to make the data input go smoothly.

Data   You'll use the perimeter you created in the last tutorial for this. If you have not completed tutorial 4–4 successfully, you can use TheLandingPerimeter feature class from the City of Oleander geodatabase in the Data folder. Also provided is a scan called The Landing—Plat Segment from the Data\Images folder.

Tools used   ArcCatalog:
   Polygon Feature Class from Lines

ArcMap:
   Georeferencing: Fit to Display
   Georeferencing: Auto Adjust
   Magnifier window
   Georeferencing: Add Control Points
   Georeferencing: Update Display
   Georeferencing: Rectify
   Proportion tool
   Copy Parallel tool
   Extend/Trim Features tool

The first steps will be to add the previously drawn perimeter of The Landing at Eden Lake plat and the scan of the area to be traced.

Set up georeferencing

1   Start ArcMap and open Tutorial 4–5.mxd. Add the feature class you created from the last tutorial, The Landing at Eden Lake, from the MyAnswers folder. If you have not successfully completed tutorial 4–4, copy the City of Oleander geodatabase from the Data folder to your MyAnswers folder. Copy the feature class LandingPerimeter in the PropertyData feature class to your MyAnswers folder and use it.

2 Next, add the scanned image of the plat called The Landing—Plat Segment from the Data\Images folder. You'll get a message box warning that the image doesn't have pyramids built. Click Yes to build them. You'll also get a warning that the image doesn't have a valid spatial reference. Click OK.

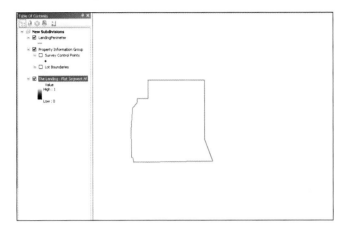

This added the necessary data to perform the georeferencing. Next, you'll add the toolbar and start making control points.

3 Add the Georeferencing toolbar from Customize > Toolbars on the main menu.

The image you need to work with has no spatial location defined, so it doesn't appear in your map display. You could zoom to it, but then you'd have trouble setting the control points. Instead, you will use one of the Georeferencing tools to move it into the current display.

4 Click the Geoprocessing drop-down arrow, and then click Fit to Display. Next, click the drop-down arrow again and clear the Auto Adjust check box. Now, you'll be able to see the scanned image and the subdivision boundary in the same view. Auto Adjust will automatically reset the transformation every time you enter a location. Since this can be confusing until you get the hang of it, it's best to turn it off for now.

Perform georeferencing

In the scan image, identify the perimeter of the subdivision. The control points will start at a location on the perimeter, and then connect to its corresponding location in the feature class representation. This needs to be done with care and as much precision as possible. To aid the process and avoid excessive zooming in and out, you'll use a magnifier window.

1 On the main menu, go to Windows > Magnifier. When the magnifier window opens, right-click the blue title line and click Update while Dragging. Then click the Magnifier drop-down arrow and set the zoom level to 600%. The process will be to drag the magnifier to a corner of the perimeter in the scan to get a close-up look at the line work. Then you will set a control point to start a link line. Next, you will drag the window over to the corresponding feature class line and set another control point at the end of the link line. The extra magnification will let you be more precise in setting the control points.

2   If you like, set the symbology of The Landing at Eden Lake layer to a thicker, red line to make it more visible. Also, start an edit session for the geodatabase containing this layer.

3   On the Georeferencing toolbar, click the Add Control Points tool.

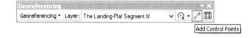

4   Drag the magnifier to the lower right of the scanned image and click one of the plat corners in the scan to set the start of the first control point.

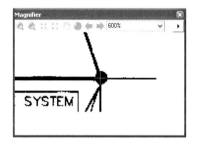

4–1
4–2
4–3
4–4
4–5

5   Drag the magnifier over the corresponding corner in the feature class, snap to the vertex, and click to end the control point.

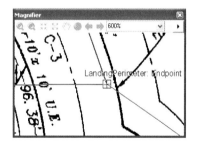

YOUR TURN

Repeat the process of setting control points for the other three extreme corners of the outline. You may, of course, do more if you desire. When you have several control points entered, close the magnifier window.

When you do the adjustment, ArcMap will stretch the scan to fit the control points as well as possible. With more control points, the transformation will be smoother.

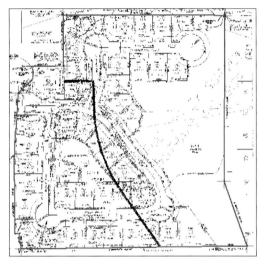

6 **Click the Georeferencing drop-down arrow, and then click Update Display.** The scan adjusted to fit the control points. The image will be altered in both scale and rotation.

If your result doesn't look right, you may have entered a pair of control points incorrectly. To fix this, click the View Link Table button on the Georeferencing toolbar. This will open a dialog box showing all your entries and the amount of correction, called the Residual, that they need to perform. The numbers should be around 1, so if you see very large numbers, you are certain to have a problem. To demonstrate what this would look like, the accompanying example has an erroneous control point pair entered. The entry is highlighted in blue in the link table and in yellow on the map. The length of the yellow line that can be seen on the map indicates a problem. The control link would need to be corrected for the image to be georeferenced correctly.

Link Table

Link	X Source	Y Source	X Map	Y Map	Residual
1	2399725.677387	6999233.274317	2399488.609857	6999546.090149	8.43833
2	2398020.207682	6999234.861785	2398460.218541	6999528.960919	11.85951
3	2397974.700268	6999345.984540	2398431.943663	6999595.145498	13.25794
4	2398357.809194	7000937.685710	2398646.650255	7000563.661277	24.72608
5	2399579.630338	7000917.048627	2399383.689463	7000563.802025	9.82560
6	2399557.405787	6999708.456382	2399381.969650	6999831.834985	2.15698
7	2398112.809978	7000562.514124	2398502.734860	7000333.548220	23.70667
8	2398124.101114	7000479.349409	2398443.783863	7000214.827985	72.77545

☑ Auto Adjust Transformation: 1st Order Polynomial (AI ⌄) Total RMS Error: [　　　]

[Load...] [Save...] [Restore From Dataset] [OK]

Save the results

Once you're satisfied that the image is adjusted to the best fit, and no more control points are necessary, you can create a new copy of the image with a spatial reference attached called a World File.

1 Click the Georeferencing drop-down arrow and select Rectify. Change the location to save to **MyAnswers**, the name to **The Landing Rectified**, and set the image type to TIFF. Then click Save. The georeferencing process will have created three new files. One is the TIFF, and the other two carry the additional information for the spatial reference. These two files have the extensions .rrd and .aux. If you move a rectified image, make sure to get all three files, or there will no longer be a spatial reference.

Save As		[?][X]
Cell Size:	0.276685	
NoData as:		
Resample Type:	Nearest Neighbor (for discrete data)	v
Output Location:	C:\ESRIPress\GIST3\MyAnswers	📂
Name:	The Landing Rectifier Format: TIFF	v
Compression Type: NONE v	Compression Quality (1-100):	75
	Save	Cancel

4-1
4-2
4-3
4-4
4-5

2 With the Georeferencing process completed, you can close the Georeferencing toolbar.

Create new features by tracing

Things get a little simpler now. All you have to do is trace the lines in the image, creating features in the feature class along the way. There are, however, some other tricks to making the new features as accurate as possible.

1 Start an edit session, making sure to edit the geodatabase that contains LandingPerimeter.

2 Select the line segment where Pintail Parkway intersects the left side of the subdivision and zoom in so that the dimensions can be read easily.

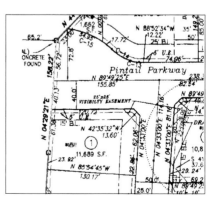

You'll use some of the COGO tools as well as standard editing tools to ensure that the lines go in at the length noted on the plat. The first is the Proportion tool. It will let you enter a distance from the start point of the line to create a node that can later be used for snapping. Use the distance shown on the plat.

3 Add the COGO toolbar to the map document and click the Proportion tool.

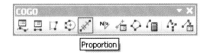

Proportion

4 The arrows show which end is the start of the line.

5 You can see from the plat that the edge of the first property line to draw starts **61.36** feet from there. Type the distance and press Enter. No other distance is required, so press Enter again to accept the remaining distance in the input screen and click OK.

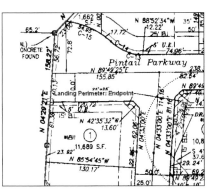

6 To begin drawing features on top of the scan image, select the LandingPerimeter layer from the Create Features template. Move the cursor over the selected feature where it was just split until the SnapTip displays LandingPerimeter: Endpoint. Click to start drawing a line.

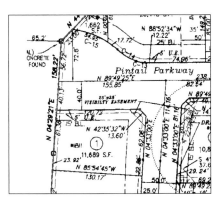

7 The length of the first line segment, according to the scan, is 120.72 feet. Right-click and use the context menu function to lock in that distance. Then move the edit sketch over the scan until it lines up with the plat and click the end of the line to establish the first segment.

8 Repeat this for the next two segments, measuring 13.60 feet and 62.06 feet, in that order.

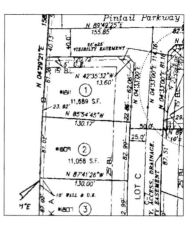

9 The next segment is an arc. Using the Endpoint Arc tool from the Feature Construction toolbar, click the endpoint and a midpoint to create the arc. The order of entry is shown on the map with red numbers.

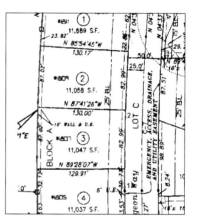

10 Pan down and enter a few more lengths from the plat. Since this is only practice, you can stop after the second arc and finish your edit sketch.

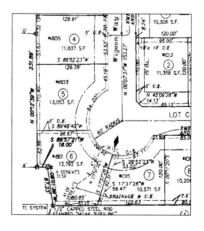

4–1

4–2

4–3

4–4

4–5

11 Go back to the left edge and select the straight arc just below the first one you selected.

12 Use the COGO Proportion tool to split it into the distances shown in the plat.

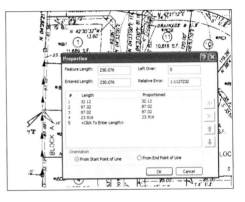

YOUR TURN

Repeat the use of the Proportion tool, using the new arc features you've drawn, to split the right side of the feature into smaller segments according to the distances shown in the scan.

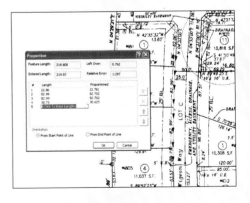

Once this is done, it's a simple matter of drawing lines between the nodes with the Straight Segment tool to create the other property lines.

13 Using the Straight Segment tool, connect the nodes to create new property lines between the new features you have created. Watch the SnapTips to verify that you are using the endpoints of the lines. By using this technique over and over, you can work your way through the plat and enter all the property lines.

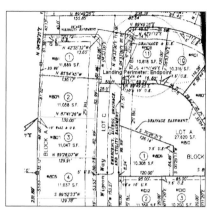

Explore another technique

There's another, simpler way to add these property line features. That is, draw the subdivision perimeter and the street edge, and then draw one property line and copy it parallel according to the distance shown on the plat. After a little cleanup work, you'll have a nice finished product.

4–1
4–2
4–3
4–4
4–5

1 Add the line along the south side of Drake Drive. Draw it as a single line segment with an arc at the end.

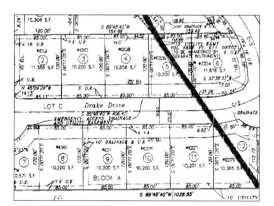

2 Now draw one of the property lines, but stop short of the edges without snapping the endpoints. This line will be perpendicular to the street edge.

3 On the Advanced Editing toolbar, select the Copy Features tool. Click the next property line to the right in the scan, and then verify that the template to draw into is correct. Click OK to complete the copy. Repeat this step for the three additional property lines.

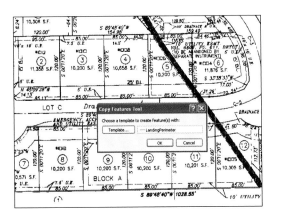

4 Select the two edges both above and below the new property lines. Then use the Extend tool on the Advanced Editing toolbar to extend them in both directions. Be sure to click both ends of each line to extend them to both of the selected features. This is a very quick way to do the tracing, especially if there are a lot of property lines that are parallel to each other. You can see that in a short time, you could have the rest of the subdivision created.

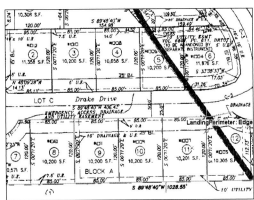

5 Save the map document to your MyAnswers folder, but keep ArcMap open.

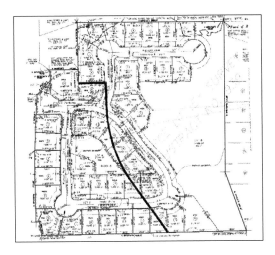

Create new polygon features in ArcCatalog

After completing the copying process, you will see that you only have lines and arcs drawn. The goal from the start, however, was to also have polygons. Retracing all these lines to make polygons would be time-consuming, but fortunately, there is a quick and easy way to create all the polygons.

1 On the right side of your map display, click the Search tab and search for the tool Create Feature Class. Click the tool to activate it.

4-1

4-2

4-3

4-4

4-5

2 Set Feature Class Location to the PropertyData feature dataset and name the feature class **TheLandingPolygons**. Set Template Feature Class to Parcels by browsing through your MyAnswers folder for the City of Oleander > Property Data feature dataset. This will make an empty feature class to store the new polygons. Note that you did not have to set a spatial reference since the feature class is going in a feature dataset with a spatial reference already set.

Now you will use the selected features to create polygons.

3 Add the Topology toolbar to your map document.

Note that even though you have not created any topologies for this dataset, you will be able to use the Create Features tool to accomplish this task. You will learn more about topology in chapter 5.

4 At this point, you may need to stop and restart your edit session in order for the new feature class to appear in the editing templates.

5 Select all the features in TheLandingPerimeter feature class. (**Hint:** Right-click and choose Selection > Select All.) On the Topology toolbar, click Construct Polygons.

6 Set the template to TheLandingPolygons and Cluster Tolerance to 0.07. Verify that Construction Options is set to Create new polygons from selected features. Click OK. The process will make a new polygon for each area enclosed by the linear features you created. Granted, it will only be a portion of the parcels in the plat, since the entire scan wasn't traced. Here is another reason why it is important that the new lines have good snapping, because anyplace where the lines did not snap will not result in an enclosed polygon.

7 When the process has completed, preview the results. With a combination of a few editing tricks and various ArcMap commands, you were able to draw the property lines with an acceptable degree of accuracy. Then in ArcCatalog, you were able to create polygons from those lines that could be merged into the parcel base to represent the new subdivision.

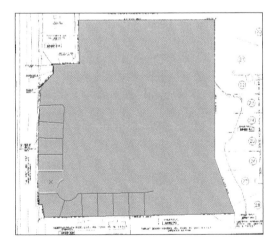

8 Save your map document as **\GIST3\MyAnswers\Tutorial 4–5.mxd**. If you are not continuing to the next exercise, exit ArcMap.

Exercise 4–5

The tutorial showed how to georeference an image and use it to add more linear features. It also demonstrated how to create polygon features from the completed lines.

In this exercise, you will repeat the process with the Running Bear Estates plat.

- Start ArcMap and open Tutorial 4–5.mxd.

- Load and zoom to the Running Bear Estates bookmark.

- Georeference the image Running Bear Estates.tif and use it to trace the new parcel lines.

- When you are done, use the lines to create a polygon feature class.

- Repeat the process with the following plats:

  - Bear Creek Estates IV

  - Creek Wood Estates

  - Little Bear II

- Save your results as **\GIST3\MyAnswers\Exercise 4–5.mxd**.

4-1
4-2
4-3
4-4
4-5

WHAT TO TURN IN

If you are working in a classroom setting with an instructor, you may be required to submit the maps you created in tutorial 4–5.

Screen captures of

         Tutorial 4–5

         Exercise 4–5

Tutorial 4–5 review

Storing scanned maps in a digital format is a common occurrence. What is less common is that these scanned maps are spatially referenced to the correct location on the ground. With no spatial reference, these digital maps are simply pretty pictures. Your organization may have thousands of scanned images, and often you will be given an older digital map and told that you need to overlay present-day features to calculate the area of new development. This exercise focused on a similar situation. The key is to create as many links as necessary to accurately place the image in the correct location. By doing this, you will create a spatial reference to the image and make it available for further analysis in GIS.

ArcMap provided the tools to accomplish this task. By using the **Georeferencing toolbar**, you were able to locate the image in real space and lock it into its real-world coordinates.

To accomplish this, you determined some common points between the **scanned image** and the **digital data** in the geodatabase. By activating the tools in the Georeferencing toolbar, you created **control points** from a point on the scanned image to the point's real location in the GIS dataset. Typically, finding these matching points by using street intersections that are both visible in the image and in the data works well, but any landmark that is common to both can be used. Since you had a plat boundary and you were georeferencing a parcel map's boundaries, you were able to match the image to existing boundary lines. Once the image was positioned correctly, you **committed those edits and saved the new image**. That saved image was then available as a reference to complete the task of adding new parcel boundaries using the COGO toolbar.

You will need at least two control points to determine the proper location, scale, and rotation of the image, but typically, you will need many more than that to match the location. As long as the control points are good and distributed evenly around the imagery, and the image is not warped or stretched too badly, the more points you can manage to enter, the more accurate the placement will be. This is an iterative process. If the first round of points does not satisfy your project specifications, you can always go back and add additional control points.

Generally when you are working on plats, you can use tools available on the **COGO toolbar** to create as much data as possible, and then use the **georeferencing tools** to locate the plat in real space and **trace** the rest of the lines. While this is not as accurate as a fully coordinated plat, it's sometimes the only way to get older data transferred.

> ### STUDY QUESTIONS
>
> **1.** How do you handle a bad control point or incorrect survey data? How do you know whether a control point is bad?
>
> **2.** What is the effect of concentrating your control points in a small area of the image?
>
> **3.** What would happen if you copied or moved the resulting image without all the associated TFW, RRD, and AUX files?

Other real-world examples

The ability to georeference imagery is a valuable tool. In some cases, being able to georeference scanned historical photographs is the only way to use older aerial imagery within GIS.

For instance, one lake management agency used a series of historical photographs to observe land-use change surrounding a dam at one of its reservoirs. Aerial imagery taken at five-year intervals from the 1960s to the present was georeferenced and overlaid with existing pertinent information contained in the geodatabase to provide a temporal view of land-use change in these locations.

A regional water supplier made use of historical as-built drawings of its pipelines to ascertain the exact locations of water distribution pipe buried in the ground. Hundreds of as-built drawings were scanned and georeferenced based on recorded fence lines, pipeline valves, roadways, and other surface features. Once georeferenced, pipe segments could be digitized from the images and related to other features in the geodatabase.

References for further study

There are a number of additional resources to help create new features in a geodatabase. **ArcGIS Desktop Help** lets you use keywords to search for a title or topic. You can access ArcGIS Desktop Help in ArcMap or ArcCatalog by clicking Help on the main menu. Then click the Search tab, type a keyword, and click Ask. Use the keywords provided here to search for additional learning resources on many of the concepts taught in this chapter.

ESRI has developed a number of online courses on a wide variety of pertinent GIS topics. **ESRI Self-Study (Virtual Campus) courses** are an excellent resource for students that will supplement information covered in this course. ESRI Virtual Campus courses can be accessed at `http://training.esri.com/gateway/index.cfm`.

4–1
4–2
4–3
4–4
4–5

ArcGIS Desktop Help search keywords

editing a geodatabase, CAD, georeferencing, Georeferencing toolbar, orthorectifying, COGO, COGO toolbar, rectify, digitize, Sketch tools, Advanced Editing toolbar, Editor toolbar, shortcut keys for editing, tips and tricks for editing

ESRI Self-Study (Virtual Campus) courses

The following ESRI Virtual Campus courses may be helpful:

1. Creating, Editing, and Managing Geodatabases for ArcGIS Desktop
2. Geoprocessing CAD Data with ArcGIS

Working with topology

Tutorial 5-1

Setting up map topology

One of the most important concepts in the field of GIS is the relationships among features. These include coincidence, adjacency, separation, and exclusion, to name a few. The geodatabase has components such as subtypes, default values, attribute domains, validation rules, and structured relationships to help construct and maintain a topological fabric. The first relationship to examine will be a map topology, which is a temporary topology built on the fly for a specific map document.

Learning objectives

- *Set up shared edits*
- *Build a map topology*
- *Edit features*
- *Create new features*

Introduction

Maintaining the integrity of your GIS data is one of the most important tasks you face. Some datasets have a natural geographic relationship that must be preserved, such as lot lines being at the edge of parcel boundaries. There may be other relationships as well, such as points representing power poles that are attached to lines representing transmission lines.

Topology, as a concept, is the idea that geographic features share a spatial relationship with each other. This may be an adjacency, an overlap, an inclusion, or a coincident edge, among other things. These relationships are critical when doing GIS analysis such as overlays or proximities. When you edit, you must be sure not to interrupt this topological relationship. If one of the features is moved, the other features that share its topology must move with it, even if the original feature is a different feature type in another layer.

One of the simplest types of topology used to maintain this relationship is the map topology. With this type of topology, you set up a temporary relationship among several layers in your map document that will allow you to do simultaneous edits of these layers. A map topology is created on the fly within a map document, and is preserved only for that map document. If the same layers are added to a different map document, the map topology will not be there, and must be created again if the same relationship is needed. There are also no user-established rules with a map topology. A basic relationship is established between the edges and nodes, which also includes the polygons they may create.

Another advantage of the map topology is that the layers you wish to include in the topology do not all have to be in the same feature dataset, which is a requirement of other topology types. The map topology may also include shapefiles, provided they are accessible in the same edit session, so that they all may be edited at the same time.

Once a map topology is created, a different toolbar is used to edit. This can allow shared edits, or even create new features to restore the topological structure.

The other two types of topology covered in this book are geometric networks, which let you set up topology rules between lines and points, and geodatabase topology, which has a rich set of rules for defining relationships among points, lines, and polygons. Both types of topology are covered in this chapter, in tutorials 5–2 and 5–3.

Setting up topological relationships

Scenario The City of Oleander recently made changes to the city limit line. Two of these changes happened through a joint annexation with adjacent cities, and one was the result of an error that was recently discovered. Because of these changes, the police districts layer is now incorrect.

5–1

5–2

5–3

There are two layers that represent the police districts. One is the polygons layer used for color shading and overlay procedures, and the other is the linear features layer used to symbolize the boundaries between districts. The linear features are in their own feature dataset and could be used in a polygon topology to preserve the relationship between the districts and their boundaries, but they cannot be included in a polygon topology with the city limit line, because it is not in the same feature dataset.

You will use two different editing techniques to make corrections to this data. The first involves shared edits. This is where you will edit one feature class, and the other feature classes that share a topological relationship will be edited at the same time. The second involves the tools to create features, a process that will restore missing features in the topological structure.

Data The map document includes the two feature classes making up the police districts, as well as the newly edited city limit line.

Tools used ArcMap:

Map topology	Split tool
Topology Edit tool	Trace tool
Topology: Show Shared Features	Topology: Construct Features

Create a map topology

1 Start ArcMap and open the map document **Tutorial 5–1.mxd.** The map document contains the new city limit line and
the features for the police districts that
you'll be correcting. Notice a few areas
where the dark green city limit line doesn't
match up with the police district boundaries.
These are the areas to fix.

As with many of the more complex editing tasks, there are some setup procedures to be
done before editing. It is important to set the snapping environment and the selectable
layers before editing the features to be sure that the correct results are achieved.

2 Perform these tasks to set up the edit session:

• Load the PD1 bookmark from the Data folder and zoom to it.

• Start an edit session.

• Make PoliceDistricts_Boundaries the only selectable layer.

• Select the police district boundary line.

If you were to use the regular editing tools to move the green vertices shown, the changes
would only affect the linear part of the police districts and not modify the polygon component of the police districts. You'd have to repeat the process to correct the polygons. While
this doesn't seem much of a task here, if you had to do this for twenty corrections for a
total of forty operations, you'd want to simplify the task.

Next you'll build a map topology that will let you edit both the linear and polygon components of the police districts simultaneously.

3 Click Customize, point to Toolbars, and
add the Topology toolbar to your map
document. Click Map Topology and
select PoliceDistricts and PoliceDistricts_
Boundaries to participate in the topology. Click OK. The map topology that you
just created will manage a topological relationship between the two layers with two
simple rules:

1. The edge of every police district polygon must have a line coincident with it.

2. Every police district boundary must have a polygon edge under it.

Using the Topology tools to edit will force ArcMap to keep the topological relationship intact. If one shared feature is modified—for instance, the police district boundary line—the corresponding feature in the other feature class will also be modified, such as the police district polygon.

Edit the topology

Using the Topology tools, you will be able to make edits on both map layers at the same time and maintain their topological relationship. This example uses two layers, but any number of layers can be included in a map topology.

1 Select the Topology Edit tool on the Topology toolbar.

2 Double-click the PoliceDistricts_ Boundaries line you want to modify, and it will turn into the edit sketch. Although this looks like the regular edit sketch, it will perform shared edits because of the tool you have selected.

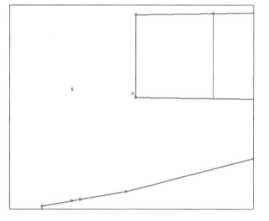

5-1

5-2

5-3

3 The Edit Vertices toolbar will display for the editing task. The default is Modify Sketch Vertices, but notice that additional tools would allow you to add or delete vertices.

4 Click the Show Shared Features button on the Topology toolbar. A display box opens showing which layers are participating in the topology. Click one of the layers and it will highlight in the map, and then click the other layer. These are the features that will be updated. Close the display box.

5 Drag the upper-left node to the right and snap it to the city limit line. Then repeat for the lower-left node. This will make the district boundary line coincident with the city limit line.

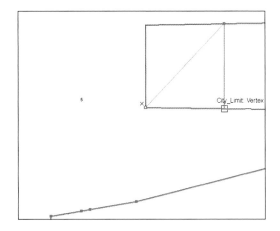

6 When you have moved both nodes, click Finish Sketch on the Edit Vertices toolbar (or press F2). Both the line and polygon layers of the police districts were modified.

This was a simple edit that required moving a few nodes. The next edit will involve creating some new features.

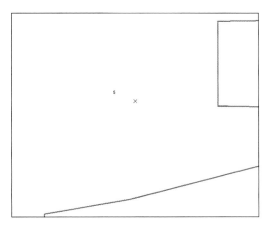

Create features using Topology tools

1 Load and zoom to the PD2 bookmark. In this region, the police district goes past the city limit line. It needs to be trimmed back, and an area needs to be removed. There is not a simple edge movement that can fix this problem, and moving nodes would take a long time. The solution will be to correct

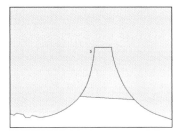

the police district boundary feature, thus making it violate the topological relationship it has with the police district polygons. Then the Topology tools can be used to correct the topological relationship, which will automatically correct the polygons to respect that relationship. Pay close attention to setting the selectable layers so that the wrong layer is not altered or deleted.

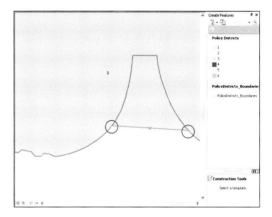

2 Start editing, if necessary, and make PoliceDistricts_Boundaries the only selectable layer. Use the Edit tool on the Editor toolbar to select the polygon boundary line in the area to correct. Then use the Split tool on the Editor toolbar to cut the line at one of the points shown in the graphic, which is where the district goes outside the city limit. Next, select the line again and split it at the other point shown. Select the resulting split line segment that crosses the unwanted area and delete it. Notice that this does nothing to change the polygon.

Deleting that section of the police district boundary caused a topology error: a polygon edge exists with no boundary line over it. The next step will be to create a polygon boundary with no coincident line. By drawing a new police district boundary, the underlying polygon will be in violation of the topology rules. Then ArcMap will fix all the associated features as it corrects the topology errors.

3 Select the PoliceDistricts_Boundaries layer from the Create Features template.

4 Starting at one end where you clipped the polygon boundary line, snap to PoliceDistricts_Boundaries: Endpoint and drag the Trace tool around the city limit line. When you reach the other endpoint, double-click to end the feature. Watch the SnapTip to ensure that you are snapping to the correct endpoint.

5–1

5–2

5–3

5 Click Split Polygons on the Topology toolbar. Click "Split existing features in target layer using selection" in the dialog box, and then click OK. Because the topology was set so that the police district polygons and boundaries must share an edge, you created a situation that broke the topology rule by reshaping the lines to follow the city limit line rather than the police district boundary. The solution to correct the topological relationship was for the Topology tool to split the polygon and create a new feature along the new edge to restore the topological structure. The new feature that was created is actually the area to be removed.

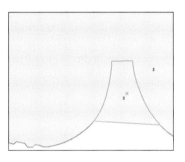

6 Change the selection to Police Districts and select the new polygon feature. Press Delete.

7 Zoom to the full map extent and examine the results. The map topology has helped to perform these edits by moving shared nodes and automatically creating features that were necessary to keep the topological structure valid.

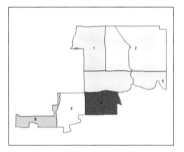

8 Save your map document as **\GIST3\MyAnswers\Tutorial 5-1.mxd**. If you are not continuing to the next exercise, exit ArcMap.

Exercise 5–1

The tutorial showed how to create a map topology and use it to make edits to the data.

In this exercise, you will complete the corrections to the police districts.

- Start ArcMap and open Tutorial 5–1.mxd.

- Load and zoom to the PD3 bookmark.

- Create a map topology for the Police Districts and PoliceDistricts_Boundaries.

- Use the Topology tools to change the police districts lines and polygons to match the new city limit line.

Hint: Construct the new boundary for the police district. Then use the Construct Features tool on the Topology toolbar to create the new extensions of the police district before you delete the old police district boundary.

- Save your results as **\GIST3\MyAnswers\Exercise 5–1.mxd**.

WHAT TO TURN IN

If you are working in a classroom setting with an instructor, you may be required to submit the maps you created in tutorial 5–1.

Printed 11-by-17-inch map or screen capture of

       Tutorial 5–1
       Exercise 5-1

5–1

5–2

5–3

Tutorial 5–1 review

In this exercise, you explored one of several types of **topology** used in ArcMap. A GIS topology is a set of rules and behaviors that model how points, lines, and polygons share geometry. Topology helps establish and maintain the relational integrity and behavior among spatial features, which is crucial for a successful GIS. A **map topology** is one of the simplest types of topology used by ArcMap and **is a temporary topology built on the fly for a specific map document**.

Topology, as a concept, is the idea that geographic features share a spatial relationship with each other. This may be an adjacency, an overlap, an inclusion, or a coincident edge, among other relationships. These relationships are critical when doing GIS analysis such as overlays or proximities. When you edit, you must be sure not to interrupt this topological relationship. If one of the features is moved, the other features that share its topology must move with it, even if the feature is a different feature type from the others and is in another layer.

Using map topology, you set up a temporary relationship between Police Districts and Police District Boundaries in your map document that allowed you to simultaneously edit these layers to match the city boundary. This preserved the **topological relationship** for the police districts, police district boundaries, and the city limit line. The topological relationship among these features helped make these edits by moving shared nodes and automatically creating features that were necessary to keep the topological structure valid.

A map topology is created on the fly within a map document and is preserved only for that map document. If the same layers are added to a different map document, the map topology will not be there and must be created again if the same relationship is needed. There are also **no user-established rules for a map topology**. A basic relationship is established between the edges and nodes, which also includes the polygons that they may create.

Another advantage of the map topology is that all the layers you wish to include in the topology don't have to be in the same feature dataset, which is a requirement of other topology types. It may also include shapefiles, provided they are accessible in the same edit session so that they all can be edited at the same time.

STUDY QUESTIONS

1. Why is topology important in a GIS?
2. List the different types of topology available in ArcMap.
3. What are the benefits of using a map topology, and in what circumstances is it best to use this type of topology?

Other real-world examples

Maintaining the integrity of your GIS data is one of the most important tasks to be aware of. Some datasets have a natural geographic relationship that must be preserved, such as shore lines being at the edge of bodies of water. There may be other relationships such as points representing train stations that are attached to lines representing train tracks.

Creating topological relationships among features is beneficial to political or jurisdictional authorities. For example, a local elections board needs to ensure that as the city boundary changes, the new areas incorporated within the city are included within the precinct. By building a topology between voter precincts and city boundaries, they can ensure that the relational integrity is preserved.

Topology is very helpful when the integrity of a network is dependent on several different features. Electric companies manage electric distribution components such as circuit segments, structures, devices, and customer and service locations. These are managed within a geometric network (see tutorial 5–2) for service and asset management needs. Power lines can be represented by edges, and structures, devices, customer locations, and service locations can be represented by junctions. Power poles have a relationship with transformers, which have a relationship with power lines. By using topology, the company can ensure that when a power pole is adjusted, the associated transformers and power lines move with it. Topological rules governing the spatial relationships help keep the electrical network functional.

5–1

5–2

5–3

Tutorial 5-2

Creating geometric networks

A second type of topology that can be managed with a GIS deals with the topological relationship between points and lines. This geometric network can control connectivity and define rules to confirm the validity of the connections. Such a network can also work in conjunction with geodatabase integrity rules to further control how data is entered and maintained.

Learning objectives

- *Build a linear network*
- *Maintain linear connectivity*
- *Establish point–line relationships*
- *Use geometric network tools*

Introduction

Geometric networks are used to build a network relationship between point and line features. This kind of topology will not only maintain connectivity for shared editing, but also allow for the creation of connectivity rules to help validate the network. The connectivity rules extend beyond the feature class level and allow unique rules to be established at the subtype level. In chapter 1, you designed subtypes to segregate data past the feature class level, and now these subtypes can be used to enforce connectivity rules.

Geometric networks are used mostly for utility networks, since there are other more appropriate tools for street networks or datasets that use linear referencing. Those must be built inside a feature dataset and can only use feature classes that do not participate in another topology. When doing your initial geodatabase design work, you will want to identify any feature classes that may be used in a geometric network and make sure that you place them in the same feature dataset.

Geometric networks are maintained as junctions and edges, which are built by the point and line feature classes you identify. Features are connected through junctions, and connectivity rules for points and lines use these junctions. ArcMap maintains a set of features and tables that make up a logical network and monitors the connectivity rules. Even though you will interact with the logical network, it is maintained automatically for you.

Junctions are defined whenever two edges connect. They facilitate the modeling of flow along the edges and through the junctions. Junctions can be rendered as enabled or disabled, controlling flow through the network.

Edges represent the path through which the items flow. They can also be rendered as enabled or disabled to control flow. Typically when a junction is defined along an edge, the edge feature is split. However, there is an option to make an edge serve as a complex edge, meaning that junctions will not cause the feature to split. As an example, if a water distribution company tracks maintenance on its pipes in 200-foot segments, splitting the features could cause problems in tracking service orders. By making the pipes a complex edge, the 200-foot segments are maintained, even though junctions are still allowed along their length.

Once built, geometric networks allow you to perform network analysis. Establishing weights for the network will assign a cost for traversing each segment in the network and more closely model reality. Weights may be derived from the material the line is made of, its size, or a combination of several factors. To establish flow in a network, you can assign sources and sinks. Sources are the points at which network flow originates, and sinks are the points at which the flow terminates. By knowing these things, ArcMap can use the logical network to assign paths or flow direction.

There are also some auto-draw actions that can be set in a geometric network. As you saw with map topology, there is a way to detect areas that break the rules of the topology and create features automatically to correct the errors. With a geometric network, you can set these rules in the connectivity matrix, and ArcMap can add the features on the fly as you draw the network.

All these elements combine within the geometric network to add behavior to your model of reality. Before designing a geometric network, it is important to study the data and take notes on how the features act in reality. Are all the line types allowed to connect to each other at every location, or is there a hierarchy of connections? For instance, the electrical service for your house doesn't connect to the high-tension line on a 60-foot tower. It connects to a local transmission line, which may go through other types of connections before reaching the high-tension lines. Likewise, you would study what point events might occur along the network. Do they represent any type of special equipment or connector necessary to make the network function? The objective is to have your computer model simulate reality wherever possible.

Setting up geometric network rules

Scenario   In chapter 4 (tutorials 4–3, 4–4, and 4–5), you completed the parcel dataset for a new subdivision in Oleander. The city's Public Works Department is interested in using GIS for its utility mapping and would like you to set up a demo using geometric networks.

5–1
5–2
5–3

The following text describes how the system behaves in reality. Remember that this is a fictitious test case and may differ from the reality in your area and that the behavior described is a typical installation. There are always exceptions to the rules that you may establish for your geometric network.

The sewer system is composed of three types of pipes. The largest are collection mains. They serve as transmission lines to take the wastewater long distances and are not used for residential services. A collection main can connect to another collection main with a manhole. None of these exists in the new subdivision, but you'll set up the rules in the geometric network anyway.

The next size is called collection laterals, and they are the lines that run parallel to the streets. When a collection lateral connects to another collection lateral, there should always be a manhole. If a line deadends, meaning that it makes no connection, there should be a manhole, although a city cleanout is sometimes placed there.

The smallest size of sewer line is the residential service. This is a private 4-inch line that connects the house to the city system. A residential service is only allowed to connect to a collection lateral, and when it does connect, it should have a tap. It may, however, connect at a manhole. The other end of the service line where it connects to the house should have a service cleanout.

Data The parcel data already exists, and an image file of the utility plan has been scanned and rectified to the parcels. In addition, a segment of the sewer lines has been highlighted for you to enter first.

Tools used ArcCatalog:

> New Geometric Network
> Geometric Network Properties

ArcMap:

> Validate Features
> Utility Network Analyst: Set Flow
> Utility Network Analyst: Display Arrows
> Utility Network Analyst: Add Junction Flag tool
> Utility Network Analyst: Trace task
> Utility Network Analyst: Solve

A geodatabase of utility data has already been created for you, but there is no geometric network built. You will build one and establish rules to match the described behavior.

Create a geometric network

1 Start ArcCatalog. Browse to the Data folder, and copy and paste the Geometric Networks geodatabase to the MyAnswers folder. Expand the copy to display the SewerNetwork feature dataset. Right-click the feature dataset, point to New, and then click Geometric Network.

2 Click Next to bypass the intro screen. Name the geometric network **SewerNetwork_ Oleander** and click Next.

5–1

5–2

5–3

3 Select the check boxes for both feature classes since both will need to participate in the network. Click Next.

4 The next dialog box allows for the feature classes to be set up as complex edges if desired, or to set sources and sinks. Leave the SewerLines feature class as a simple edge, but set the Sources & Sinks column to Yes for the SewerConnectors feature class. Click Next.

5 This particular network won't have weights, although it is possible to include them as a cost to traverse each pipe segment when doing flow analysis. Click Next.

6 The final screen is a summary of all your selections. If any of them does not appear correct, use the Back button to return to that screen and make corrections. Otherwise, click Finish. The geometric network is built, creating two new feature classes to manage the logical network. The first manages the edges and is called SewerNetwork_Oleander. The second manages the junctions and is called SewerNetwork_Oleander_Junctions.

Name	Type
SewerConnectors	Personal Geodatabase Feature Class
SewerLines	Personal Geodatabase Feature Class
SewerNetwork_Oleander	Personal Geodatabase Geometric Network
SewerNetwork_Oleander_Junctions	Personal Geodatabase Feature Class

Set up the network rules

Previously, the engineers provided a description of the network's behavior. Now you'll use this set of rules to set up the connectivity matrix. A good way to go through all the possibilities and avoid too much confusion is to rank the components by a hierarchy and set their connectivity from lowest to highest. This prevents having to backtrack through the lists and potentially alter a setting that has already been made. The hierarchy of these components is Main > Collection Lateral > Residential Service. The connectivity rules only need to be set up for the linear features, since point features cannot connect to other points.

1 In ArcCatalog, right-click SewerNetwork_Oleander, and then click Properties.

2 Click the Connectivity tab and set the
Connectivity rules for (feature class)
drop-down list to SewerLines.

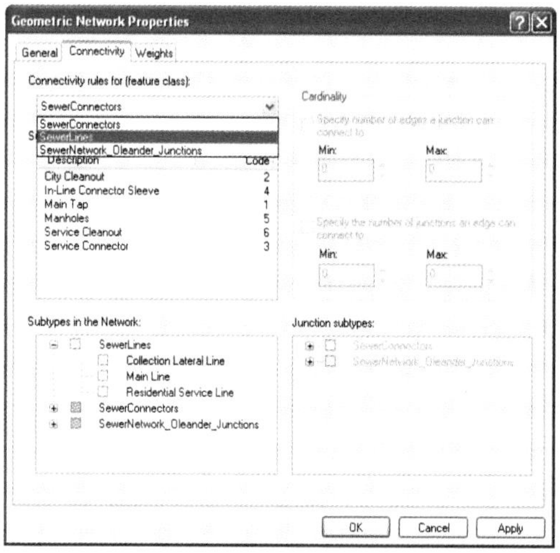

3 In the Subtypes in this feature class pane,
select Residential Service Line. Notice that
the connectivity rules are set with respect
to each subtype. This is yet another benefit of defining subtypes in your geodatabase
design. The geometric network can control
how each subtype interacts in this type of
network style topology.

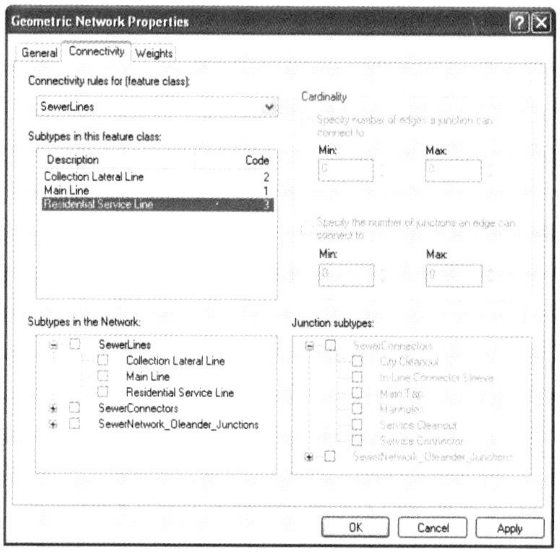

In the network connectivity matrix dialog
box, there are two input areas at the bottom.
The left network pane is used to select a feature type to which the selected feature type
will connect. The right network pane is used
to define what connections between those
features are allowed. You can also designate a default feature to be assigned when the connection takes place. The left network pane can also be used to assign a default feature when
a connection doesn't take place—i.e., the free-end case.

5–1

5–2

5–3

4 In the left network pane, click the toggle key to expand the list under Sewer Lines, if necessary. A residential service line is not allowed to connect to another residential service line, so don't select that check box. A residential service line is not allowed to connect to a main, so don't select that check box either. A residential service line is allowed to connect to a collection lateral, so select that check box. Note that the right network pane becomes active.

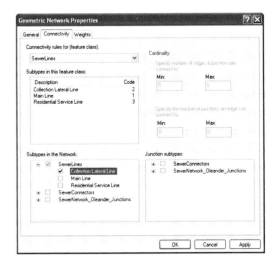

5 Expand the choices under SewerConnectors in the right network pane. When a residential service line connects to a collection lateral, it should have a service connector. Select the Service Connector check box. Notice that a blue circle with a D inside is displayed. This connection is marked as the default, so the point feature for Service Connector will automatically be drawn every time this type of connection takes place. The indicator for the default feature can be changed to another feature, but only one feature can be set as the default.

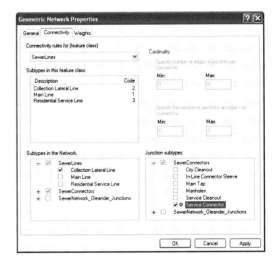

6 A connection to a manhole is also allowed, so select the Manholes check box, but don't set it as the default. Also, select the SewerNetwork_Oleander_Junctions check box to create the feature used by the logical network to manage flow analysis. Now click Apply. The rule you set up says that a residential service line is allowed to connect to a sewer lateral at a manhole or at a residential service connector. If no manhole exists at the point of connection when the line is drawn, a service connector will automatically be drawn by default.

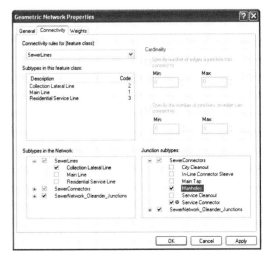

7 In the left network pane, expand the choices under SewerConnectors. Select the Service Cleanout check box. Now right-click Service Cleanout, and then click Set as Default. A small blue circle with a D inside will be added next to the name. Click Apply. The addition to the rules for the residential service line now says that if you draw a new residential service line, the geometric network should look at the endpoints you have created. If one of the ends snaps to an existing feature, ArcMap will use the connectivity rules to validate the features. If an end does not connect to an existing linear feature, ArcMap will automatically draw a service cleanout there as a result of you having made it the default feature.

This is another type of data integrity rule. If a situation exists that must always happen when new features are drawn, ArcMap will automatically draw them according to the geometric network rules.

Having set up all possible connection types for residential service lines, you won't have to visit this selection again. Notice that some of the other check boxes for other feature types were automatically selected as you went through the setup. Don't clear these, because they are being managed by the logical network. Moving on, you will set up the geometric network rules for the other features.

8 In the Subtypes selection pane, click Collection Lateral Line. You'll notice that some of the boxes in the different panes are already checked. Do not clear the check marks as they represent other rules previously set.

9 Service lines are allowed to connect to each other, so in the left network pane, select the Collection Lateral Line check box. In the right network pane, select the Manholes check box since this type of connection should result in a manhole being automatically drawn. If the blue circle doesn't appear, right-click Manholes, and then click Set as Default. When a residential service line connects to the collection lateral line, it will break the line and place a service connector at the junction. Select the Service Connector check box, so that this three-way junction will be valid. Also select the SewerNetwork_Oleander_Junctions check box. Click Apply.

10 Now consider the next step in the hierarchy. A collection lateral is allowed to connect to the main, so in the left network pane, select the Main Line check box. When it connects, it should get a main tap, so select the Main Tap check box in the right network pane and set it as the default. Also select the SewerNetwork_Oleander_Junctions check box. Click Apply.

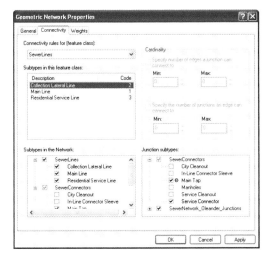

11 The default free-end case for the collection lateral is a manhole, although a city cleanout is sometimes placed there. In the left network pane, Manholes should already be selected as a result of a previous rule. Set it as the default, and then select the City Cleanout check box to make it a valid choice. Click Apply. That completes all the possible connections for Collection Lateral Line, so move up the hierarchy to Main Line.

Since everything below Main Line is already set, the only connection to consider is a main line to another main line. Remember not to clear any check boxes as you move through the menus. These are set by previous rules.

12 Select Main Line in the Subtypes pane. In the left network pane, select the Main Line check box, and then in the right network pane, select the In-Line Connector Sleeve check box, which should be set as the default feature for this type of junction. Also select the SewerNetwork_Oleander_Junctions check box. Click Apply.

13 The free-end case for the main line is a manhole. Select Manholes in the left network pane and set it as the default. Click Apply, and then click OK. Close ArcCatalog. This completes the setup for the geometric network rules. Remember that these rules would not affect any existing features in the geometric network feature classes, but they will be applied as new features are added.

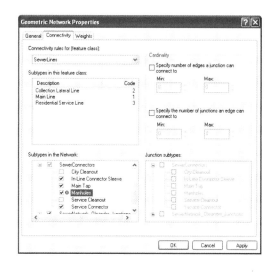

Create new network features

When creating features within a geometric network, it is important to be aware of what connections are taking place, and what features the network may automatically draw. For instance, if you were to draw all the residential service lines first, there would be no collection lateral to snap to, thus making both ends of the line free-end cases. Both would automatically have a service cleanout drawn. If the collection lateral already existed and the residential service were snapped to it, the connected end would get a service connector drawn, and the free end would get a service cleanout drawn.

The best methodology to ensure that all the connections take place as designed is to start at the top of the hierarchy and draw those features first. Then move down through the hierarchy, drawing the residential service lines last.

1 Start ArcMap and open Tutorial 5–2.mxd.

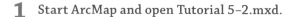

5–1

5–2

5–3

2 Right-click the Layers data frame and add a new group layer. Change its name to **Network Group**. Right-click Network Group, and then click Add Data. Browse to the MyAnswers folder and expand the Geometric Networks geodatabase and the SewerNetwork feature dataset. Click SewerNetwork_Oleander, and then click Add. This will add all the layers that participate in the topology to the group layer.

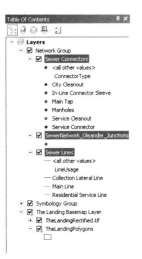

The Symbology Group in the table of contents holds some LYR files that have been symbolized for the sewer network. Use these to import symbology for your new layers.

3 Click Properties for the Sewer Connectors layer. Click the Symbology tab, and then click Import. Set Layer to Sewer Connectors and click OK. Verify the Value Field as Connector Type and click OK. Close the Properties dialog box by clicking OK.

4 Repeat the process to import symbology for Sewer Lines. Verify the Value Field as LineUsage and click OK.

5 Make sure the Sewer Connectors and Sewer Lines layers are turned on. Turn off the SewerNetwork_Oleander_Junctions layer.

6 Set up an edit session with the following steps:

- Load and zoom to the Sewer1 bookmark in the Data\Bookmarks folder.
- Start an edit session for the Geometric Networks geodatabase.
- Click Collection Lateral Line in the Create Features template.
- Select the Straight Segment tool on the Editor toolbar.

To help you identify which lines on the plans are collection lateral lines, they have been highlighted in yellow. The residential service lines have been highlighted in red. Starting at the upper part of the drawing, you will trace the lines highlighted in yellow. There is a manhole at each corner, so you will need to end the edit sketch at each corner. If the geometric network is set up correctly, the manholes will automatically be drawn.

7   Starting at the upper part of the drawing, use the Straight Segment tool to start a new edit sketch at the end of the sewer line. Move to the first manhole, shown as a black circle, and double-click to end the sketch. The line and two manholes will be drawn.

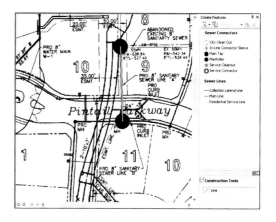

YOUR TURN

Continue tracing the sewer line highlighted in yellow, being sure to finish the sketch at each manhole. Then snap to the manhole to restart the sketch. Pan and zoom to follow the line. You will use one of the arc tools to draw the curved parts of the sewer line. When you have reached the end, turn off the image and zoom out to examine your results. If you have trouble identifying the sewer line, match the final results in the accompanying figure to the image and see if you can follow it across the drawing.

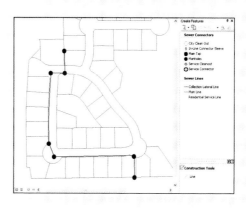

5-1

5-2

5-3

Notice how the geometric network automatically drew the manholes. If you were to look at the attribute table of the Sewer Connectors layer, you would see a number of features there, even though you never had to change the drawing template or actually draw them.

Next you will go back and add the residential service lines.

8 Return to the Sewer1 bookmark and turn the image layer back on. Click Residential Service Line in the Create Features template and select the Straight Segment tool. Zoom back to the Sewer1 bookmark if necessary. Locate the first residential service in the image and click to start at the upper end. Then move down until the cursor snaps to the collection line and double-click. The appropriate features are automatically drawn at each end and symbolized accordingly. A residential service

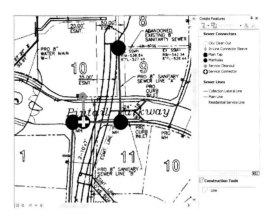

cleanout point feature was added at the free end, and a service connector was added at the end that snapped to the existing collector. If snapping was not on or you were not within the snap tolerance when you ended the line, the connector feature would not have been drawn, so always monitor the snapping environment when working in a geometric network.

YOUR TURN

Move around the image and draw the rest of the highlighted residential service lines. Try drawing the points in reverse order and see if it makes any difference to the auto-draw function. You can take artistic license and draw the service lines longer than they appear so that the symbols display clearly. When you have completed the sketch, turn off the image and zoom out to see your work.

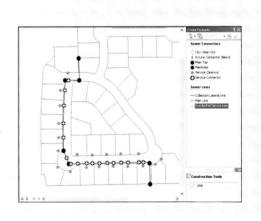

Validate features

One of the bonuses of establishing connectivity rules is that ArcMap can identify places where the rules have been broken.

1 **Using the Select Features tool, drag a box over the entire map to select all the features. On the Editor drop-down menu, click Validate Features.** The logical network is being tested against the geometric network rules that you established. All the connections between line styles are examined and any violations are reported. By using the auto-draw feature of the geometric network, you can be assured that none of the point features is invalid.

2 **Two features were reported as invalid. Select one of them and click Validate Features again. After reading the message, click OK.** In a long cryptic statement, you are told that the residential service line is not allowed to connect to a manhole. The reality of the situation is that in this rare case, the residential services were connected directly to the manhole, so you can ignore the error as an exception.

5–1
5–2
5–3

Establish flow

Although geometric networks are not intended to provide complex analysis, simple routing and flow analysis can be done. The more sophisticated networks that can be built with the Network Analyst extension include a more robust weight system, support for turn impedance, better control over connectivity, multimodel models, and allocation operations.

The flow in a geometric network is based on either the source of flow or a sink designating the destination of flow. In any one network, only one type is defined. For this example, you'll place a sink at the destination point of the lines and let the logical network calculate how the flow will reach the sink.

The manhole at the upper part of the drawing (farthest north) is the low point in this sewer system and will be designated as the sink.

1 Zoom to the Sewer1 bookmark. Set Sewer Connectors as the only selectable layer and select the northernmost manhole as shown in the accompanying figure. On the Editor toolbar, click the Attributes button. Set the field AncillaryRole to Sink. Close the Attributes dialog box. Setting the ancillary role established how the feature will function in the network. Making a feature a sink will cause it to attract flow. Setting it as a source will cause the flow to originate from there. Use only one ancillary type in a network, or the flow will work against itself and the logical network won't be able to determine the direction of the flow.

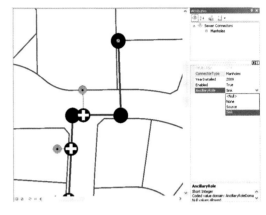

2 Add the Utility Network Analyst toolbar to your map document. Click the Set Flow Direction button to establish flow direction.

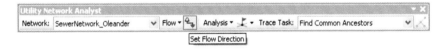

3 Click the Flow drop-down arrow and select Properties. Click the Arrow Symbol tab and make sure Determinate flow is selected. Next click the displayed symbol. When the Symbol Selector dialog box opens, set the size to 7. Click OK, and then OK again.

4 Click the Flow drop-down arrow again and select Display Arrows. Zoom to the full extent to see the results. The arrows will display the flow direction, which is useful for a network such as a sewer system or stream system. Networks such as streets or water don't have a single direction of flow, but they do provide a mechanism for calculating routes.

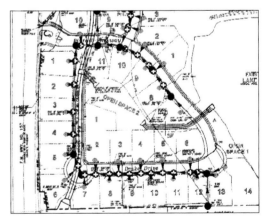

Perform network tracing

Geometric networks also allow for simple path analysis. This may be routing through streets, or in the case of a sewer system, a tracing of flow.

1 On the Utility Network Analyst toolbar, click the Flags tools drop-down arrow and select the Add Junction Flag tool.

2 Move the flag cursor over and click one of the residential service cleanout nodes. A large green box will appear to mark the flag location.

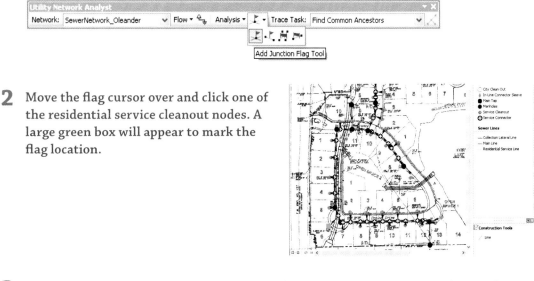

3 Set Trace Task to Trace Downstream. Then click the Solve button. The path from the selected service to the sink is shown in red. Can you tell how many linear feet of pipe this represents? No, because the features are not selected, only highlighted.

5-1

5-2

5-3

4 Click the Analysis drop-down arrow, and then click Options. Click the Results tab, and in the Results format pane, click Selection. Take a moment to look at the other tabs and options before closing the dialog box.

5 Make the Sewer Lines layer selectable, then solve for Trace Downstream again, and notice that the path features are selected. Now could you get the total linear feet of pipe? Yes, by looking at the selected records in the attribute table.

6 Stop editing and save your edits.

Establish network weights

The sewer network shows how tracing along the network can be accomplished, but sewer networks always flow the same way—downhill. To see how paths can be traced in a multiflow environment, you'll add a water utility network to the map document. The network also has weights established to provide a more interesting investigation.

1 Add a new group layer to the map document and name it **Water Utility Group**. Then add the WaterDistribution feature dataset from the Geometric Networks geodatabase to the new group.

2 Load and zoom to the Water1 book-
mark, and turn off all the layers except
DistLateral, DistMains, and hydrants.
Start a new edit session.

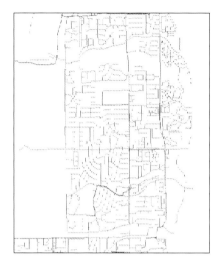

3 On the Utility Network Analyst toolbar, change Network to WaterDistribution_Net
and Trace task to Find Path. A network is already established for this dataset consisting of
three different linear feature classes. The network can trace paths across the three feature
classes, and with no weights set, it will look for the shortest path. You will set a few flags on
the network to see how it functions.

4 Select the Add Junction Flag tool and place
two junctions on the network—one near
the upper right corner of the city and one
near the lower left corner of the city. The
choices used for the tutorial are marked
with a red circle, but you may choose any
location you like.

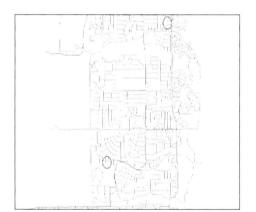

5–1
5–2
5–3

5 Next, click the Solve button on the Utility Network Analyst toolbar and note the path that is selected. You may need to verify that the path is being returned as a set of selected features.

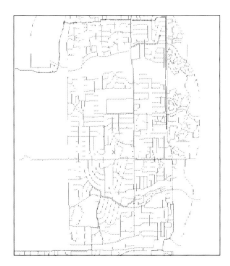

6 Open the attribute tables for both the DistLateral and DistMains layers. Get totals of the selected features for the fields Cost in time and Shape_Length for each layer and add them together for a total cost in minutes and a total length. Note that both tables can be displayed in the Table window, and you can move between them using the tabs at the bottom of the dialog box. When you have your results, close the Table window.

Shape *	PSIZE	PTYPE	Enabled	year_const	XDList	Shape_Length	Cost in time
Polyline	12"	Polyvinyl Chloride	True	2005	<Null>	140.008035	0.583367
Polyline	6"	Cast Iron	True	1965	<Null>	112.004465	0.933371
Polyline	6"	Cast Iron	True	1965	<Null>	390.149196	3.251243
Polyline	6"	Cast Iron	True	1965	<Null>	603.31469	5.027622
Polyline	6"	Cast Iron	True	1965	<Null>	380.529616	3.17108
Polyline	6"	Cast Iron	True	1965	<Null>	347.401194	2.89501
Polyline	6"	Cast Iron	True	1965	<Null>	83.899073	0.699159
Polyline	6"	Cast Iron	True	1965	<Null>	47.513963	0.39595
Polyline	6"	Cast Iron	True	1965	<Null>	308.876353	2.57397
Polyline	6"	Cast Iron	True	1965	<Null>	425.057253	3.542144
Polyline	6"	Cast Iron	True	1965	<Null>	572.560881	4.771341
Polyline	6"	Polyvinyl Chloride	True	1993	<Null>	25.501193	0.21251
Polyline	6"	Polyvinyl Chloride	True	1993	<Null>	106.014445	0.883454
Polyline	6"	Polyvinyl Chloride	True	1993	<Null>	33.508427	0.279237
Polyline	6"	Cast Iron	True	1965	<Null>	567.027905	4.725233

(0 out of *2000 Selected)

DistLateral

Next, you'll use a set of weights developed for this dataset that will take into account the size of the pipes when determining the best path. The field Cost in time takes into account the size and length of each pipe segment to calculate how long it takes water to flow from one end to the other. By using this as the weight field, the calculated path will be the fastest, not the shortest as calculated earlier.

7 On the Utility Network Analyst toolbar, go to Analysis > Options and click the Weights tab. Set both the Edge weights fields to FlowRate. Click OK.

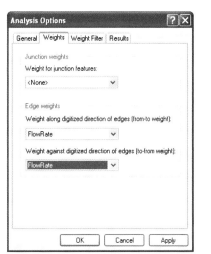

8 Click the Solve button again and note that a different path is taken. Once again, get the totals for Cost in time and Shape_Length and compare with the previous totals. When you are done, stop the edit session.

You can see that using the weights increased the overall length of the traced path, but with the use of larger pipes, the trip is made in a shorter time.

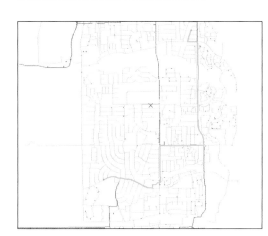

5–1

5–2

5–3

Although geometric networks don't have the robust functionality of the Utility Network Analyst extension, they can still provide useful analysis tools right out of the box.

Set up shared edits

Another important advantage to having your data in a geometric network is that the connectivity is maintained by the logical network during editing. When features are edited, all the features that have a topological relationship with that feature will be moved at the same time.

The sewer lines that you drew represent the initial plan for construction, but now you discover that two manholes had to be relocated to avoid an underground pipeline.

1 Make sure the Network Group and The Landing Basemap Layer are turned on. Load and zoom to the Sewer2 bookmark and start an edit session. The Public Works Department has reported an error in the original drawings, and now the two manholes on Pintail Parkway need to be moved to the center of the street. Make the Sewer Lines layer selectable and select the two segments of the collection lateral lines that run along Pintail Parkway.

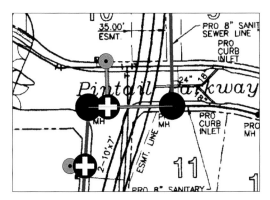

2 Using the Edit tool from the Editor toolbar, drag the two lines to the center of the street. Note that all the associated manholes, service lines, and service connectors move, too. Turn off the image file and examine the results. It is important to note that edits to the geometric network can be done with the standard editing tools, and the logical network is still maintained. If you wish, experiment with moving various lines and nodes to see how they are controlled by the network.

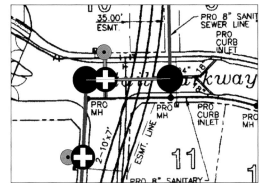

3 Save your map document as **\GIST3\MyAnswers\Tutorial 5–2.mxd**. If you are not continuing to the next exercise, exit ArcMap.

Exercise 5–2

The tutorial showed how to edit features within the confines of a geometric network. In addition, it demonstrated some network analysis tools.

In this exercise, you will add the remainder of the sewer lines for this subdivision, using all the geometric network rules.

- Start ArcMap and open Tutorial 5–2.mxd.

- Add the image file GeometricNetworkCompleteRectify.tif from the Data\Images folder.

- As with the tutorial, the collection lateral lines are highlighted in yellow and the residential service lines are highlighted in red.

- Construct the remaining sewer lines to complete the subdivision.

- Establish the flow direction and demonstrate the use of the path tracing tools. Leave the traced features selected, so that they will be visible on the printed map or screen capture.

- Save your results as **\GIST3\MyAnswers\Exercise 5–2.mxd**.

5–1

5–2

5–3

WHAT TO TURN IN

If you are working in a classroom setting with an instructor, you may be required to submit the maps you created in tutorial 5–2.

Printed 11-by-17-inch map or screen capture of

Tutorial 5–2

Exercise 5–2

Tutorial 5–2 review

Geometric networks are used to build a network relationship between point and line features. The network not only will maintain connectivity for shared editing, but it also allows for the creation of connectivity rules to help validate the network. A **geometric network** must be built inside a feature dataset and use feature classes that are not participating in another topology.

You developed a geometric network for the sewer lines for Oleander using the **Build Geometric Network wizard** in ArcMap. The Build Geometric Network wizard guided you through all the steps necessary to create your geometric network. Prior to creating this geometric network, ArcMap was not aware of how features in the same feature class or features in different feature classes related to each other. By using the Build Geometric Network wizard, you specified **connectivity rules** for how the main lines, collection laterals, and residential service lines related to each other in the SewerLines feature class, as well as how these features were related to the manhole and cleanout features contained in the SewerConnectors feature class.

Geometric networks facilitate the modeling of flow along two special feature types of points and lines: **junctions** and **edges**. Rules pertaining to the junctions and edges were created that helped characterize the flow within the geometric network. In the exercise, the junctions were manholes and cleanouts, and the edges were the sewer lines. The sewer lines were connected to each other through the manholes, and the cleanouts were junctions at the end of each sewer line.

Once built, the geometric network allowed you to perform **network analysis**. You did this by specifying **sources** and **sinks** for your junctions. Sources are the points at which network flow originates (in the exercise, these were the manholes), and sinks are the points at which the flow terminates (in the exercise, these were the cleanouts). By specifying these parameters, ArcMap was able to use the logical network to assign paths, or flow direction.

You also used **auto-draw** actions that you set in the **connectivity matrix** within the Geometric Network menu. With these rules set, ArcMap added features on the fly as you drew the network. The connectivity is **continuously updated** whenever any network feature is added or removed from the geometric network.

STUDY QUESTIONS

1. When you established your geometric network rules, you established a hierarchy from largest to smallest sewer lines. Why was it important to create these features in the same order?

2. With respect to data organization within feature classes and feature datasets, how did the geometric network differ from map topology? Give an example.

3. What is the importance of designating sources and sinks within a geometric network? Give examples of when sources and sinks should and should not be used in your geometric network.

4. Name two advantages of incorporating a geometric network in your geodatabase.

Other real-world examples

As illustrated in the exercise, a geometric network can be developed within ArcMap that can model a real-world system. To create the geometric network, however, you need to understand how each feature operates within the network. During the development of your geodatabase, it is best to sit down with the industry experts and derive exactly how this system operates. Once you understand how the system functions, you can translate its operational characteristics using the geometric network tools in ArcMap and build them into your geometric network. While doing the initial geodatabase design work, you will want to identify any feature classes that may be used in a geometric network and make sure that you place them in the same feature dataset.

For example, a water distribution company tracks maintenance on its pipes between major appurtenances (such as maintenance valves and manholes) along the pipeline. At each appurtenance, a junction is placed to denote the corresponding line between each pair of appurtenances. Edges are used to represent the pipeline, and junctions represent each appurtenance. By placing these junctions, the geometry of the pipeline is modified so that work orders can be attached to each segment.

Gas companies can manage their gas distribution lines using a geometric network. Pipelines are represented by edges, and valves, taps, and tees are represented by junctions that can be opened and closed. Tracing procedures can be used to see which pipelines are connected to each other or which areas are under the same pressure regulation, or even to simulate product flow.

5-1
5-2
5-3

Tutorial 5–3

Setting up geodatabase topology

The most robust of the three types of topology, geodatabase topology, is what most people are referring to when they use the term topology. Geodatabase topology allows for the user to establish which rules to enforce and supplies a rich set of tools for examining and correcting topology errors.

Learning objectives

- *Establish topology rules*
- *Build a geodatabase topology*
- *Examine topology errors*
- *Correct topology errors*

Introduction

You learned that map topology was useful for creating a temporary, limited topology within a single map document. You also learned that a geometric network is useful when creating and editing a utility network. The last level of topology, the most sophisticated of the bunch, is geodatabase topology. This type of topology is constructed against existing feature classes and stored in the geodatabase for future use. Unlike map topology, geodatabase topology persists beyond the map document.

Geodatabase topology can model the behavior of points, lines, and polygons. There are also rules for like features, such as polygon-to-polygon behavior. ArcGIS contains a set of predefined rules to manage these relationships, as illustrated on the ArcGIS Geodatabase Topology Rules poster. This can be found in the Documentation folder where your ArcGIS is installed, or from the student DVD included with this book. The point rules are shown in purple, the line rules in green, and the polygon rules in blue.

The first step in building topology is to determine what relationships exist in your data. Think about a street network as an example. One rule is that each street segment should be drawn only once, and it should not have a bunch of lines overlapping. A review of the topology rules poster shows two linear rules stating that lines Must Not Overlap and lines Must Not Self-Overlap. Read the descriptions of these rules to determine if either or both might be useful in your application. Roads also must not go down rivers, so perhaps the rule Must Not Overlap could be applied to streets and rivers.

After a set of rules is determined for the topology, you must gather all the feature classes that will participate in the topology and place them in a single feature dataset. There are two important rules to remember:

1. All feature classes in a topology must be in the same feature dataset.

2. Feature classes can only participate in one topology (any that can be created in a feature dataset).

Finally, the topology can be built. As stated before, geodatabase topologies are saved as an independent item in the feature dataset and will persist across many map documents. They can also be updated later to add or subtract rules or layers.

When the topology is built and validated, a new set of feature classes is created. These hold indicators to all the features that violate the rules you have established and are used to make corrections. Among the options to correct the topology errors are to manually fix the error and revalidate the topology, use the Topology tools to perform shared edits on the features, or use the topology corrections procedures to automatically fix the errors. The severity of the error and the features involved will dictate which method to be used.

Setting up geodatabase topology rules

Scenario   You worked with the City of Oleander parcel data in previous tutorials to design the geodatabase and do some editing. In the future, you may be the one to maintain this data, so you will need to establish a set of topology rules that can help with the task.

Data   Examine the feature classes in the Property Data feature dataset in the City of Oleander geodatabase. Refer to the topology rules poster to select rules to use on the following features:

> Blocks = polygons representing sets of parcels for a particular block in a subdivision
> City area = polygon representing the city limits
> City boundary = line representing the city limits
> Lot boundaries = lines representing the parcel boundaries
> Parcels = polygons representing pieces of property
> Subdivisions = polygons representing platted subdivisions

Tools used   ArcCatalog:

> New Topology

ArcMap:

Topology: Error Inspector	Topology: Subtract
Extend tool	Topology: Construct Features
Trim tool	Topology: Fix Topology Error tool
Topology: Validate Topology	

5–1
5–2
5–3

Select topology rules

1 Open the topology rules poster from the provided DVD. Starting at the upper left of the diagram, read the rules and evaluate how each rule applies to the land records data. Think also of how editing one of the feature classes might affect another.

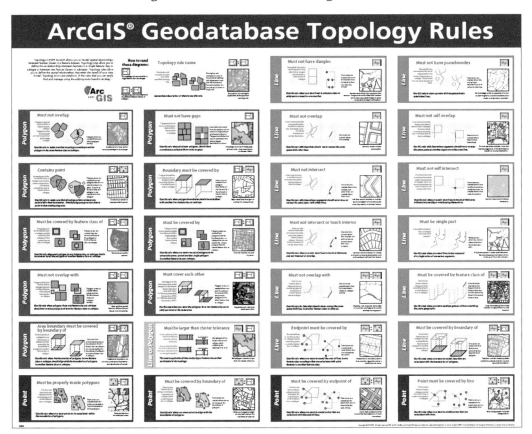

Polygons:

- **Must Not Overlap:** For the feature class Parcels, you can't have overlapping polygons or it would mean that two people own the same piece of property. Other features that are mutually exclusive are Blocks and Subdivisions.

- **Must Be Covered by Feature Class of:** All the polygon-type feature classes must be covered by the feature class CityArea. Any features that fall outside this area are not in Oleander.

- **Must Not Have Gaps:** The parcels must not have gaps between them, or there would be property that no one owns. Since blocks and subdivisions occur only on platted property, they are allowed to have gaps.

- **Boundary Must Be Covered by:** The perimeter of the CityArea polygon features must be covered by CityBoundary linear features. Use this rule when you intend to edit the polygons and have the line work automatically adjust. Also, the blocks and subdivisions should be covered by a lot boundary of the platted property. This rule is also used when there is no one-to-one relationship between the polygons and the line work.

Lines:

- **Must Not Have Dangles:** The line features in both the LotBoundaries and CityBoundary feature classes must enclose polygons, so they cannot have lines with ends that do not snap to another line.

- **Must Not Overlap:** The LotBoundaries feature class should not have multiple features representing the same lot line. This creates a problem both with editing and symbolizing. This rule should also apply to the CityBoundary feature class.

- **Must Not Self-Overlap:** It is never desirable to have a line feature double back on itself. Set this rule for both the LotBoundaries and CityBoundary feature classes.

- **Must Be Covered by Boundary of:** The lot boundaries must be covered by the edge of a parcel feature. The intent here is to edit the line work and let the polygons be automatically created. Also, there is a one-to-one relationship between the lot boundaries and the parcel edges.

Note that some of these rules apply only to a single feature class, while others establish a relationship between two feature classes, sometimes of different feature types. In your evaluation of the rules, you may have decided to add a few not listed here. This will make the rules more stringent, but be careful not to apply two rules that either do the same thing or conflict with each other.

5–1

5–2

5–3

Build the topology

1 Start ArcCatalog and browse to the Data folder. Copy the City of Oleander geodatabase to the MyAnswers folder. Expand the geodatabase and review the Property Data feature dataset.

2 Right-click the Property Data feature class, point to New, and then click Topology. Click Next to bypass the intro screen.

3 Accept the default name. The existing data will need to be run through a snapping process to build the relationships among the existing geographic features. To perform a more vigorous snap, remove two zeros to the right of the decimal place in the cluster tolerance. Click Next.

4 Select all the layers except LandingPerimeter and ZoningDistrictsMasks to participate in the topology. Click Next. In the topology, each layer must have a ranking, where 1 is highest. The layers with the highest rank will not be altered in a snapping process. Layers with lower ranks will be adjusted to be coincident with the higher ranked layers.

In this example, the lot lines and city limit boundary are the highest ranked since they are typically entered from accurate survey data. Next will be the parcels and city limit area, followed by the blocks and subdivisions.

5 Set the number of ranks to **5**. Click the rank next to Parcels and change it to **2**. Set the rank for City Area to **2**, Blocks to **4**, and Subdivisions to **5**. Click Next.

6 In this dialog box, you will set up the topology rules. Click Add Rule, and then in the Features of feature class list, select Parcels. Set the rule to Must Not Overlap, and click OK. The rule will be added to the topology. This rule basically states that any piece of property cannot overlap another piece of property. Otherwise, two people would unknowingly own the same piece of land.

Notice that the topology rule can be set to a feature class, or to a subtype of a feature class, similar to the way subtypes can be used in geometric networks. This may be an important consideration during the design phase of a geodatabase.

5–1

5–2

5–3

YOUR TURN

In the topology rules dialog box, use Add Rule to set the following rules:

Feature class	Rule	Feature class
Parcels	Must Not Overlap	
Blocks	Must Not Overlap	
Subdivisions	Must Not Overlap	
Parcels	Must Be Covered by	CityArea
Blocks	Must Be Covered by	CityArea
Subdivisions	Must Be Covered by	CityArea
Parcels	Must Not Have Gaps	
CityArea	Boundary Must Be Covered by…	CityBoundary
Blocks	Area Boundary Must Be Covered by …	Parcels: Platted
Subdivisions	Area Boundary Must Be Covered by…	Parcels: Platted
LotBoundaries	Must Not Have Dangles	
CityBoundary	Must Not Have Dangles	
LotBoundaries	Must Not Overlap	
CityBoundary	Must Not Overlap	
LotBoundaries	Must Not Self-Overlap	
CityBoundary	Must Not Self-Overlap	
LotBoundaries	Must Be Covered by Boundary of…	Parcels

Finish the topology creation and click Yes to validate it.

Match some of these rules to the topology poster and see if you can understand what each is doing.

The validation process will check the existing data against the new topology rules you created. This first time through, there will be a lot of topology errors that need to be fixed. In the future, you would only have to fix topology errors after editing.

7 Click Property_Data_Topology to preview the results of the topology validation. Zoom and pan to see the extent of the errors, and then close ArcCatalog.

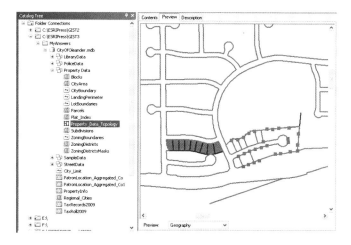

Examine the linear errors

5-1
5-2
5-3

When the Property_Data_Topology is brought into ArcMap, you will be able to examine the errors and make corrections. The goal will be either to fix every error or to mark them as exceptions to the rule.

1 Start ArcMap and create a new, empty map document. Add only the Property_ Data_Topology layer to the map, and when prompted, accept the addition of all the feature classes that participate in the topology. Examine the resulting table of contents. All the feature classes that have topology rules established are added to the TOC, along with a layer representing all the topology errors. This layer contains points, lines, and polygons (shown as areas) and is symbolized.

The number of errors may look overwhelming, but by taking them in small groups, you can tackle them. By using the topology properties, you will restrict which errors are shown.

2 Start an edit session and add the Topology toolbar to the map document. Click the Error Inspector button and move the resulting dialog box so that you can see the dialog box, the Topology toolbar, and the topology errors all at the same time. Remember that you can prevent the dialog box from docking to the edge of the map by pressing Ctrl while moving the box.

3 Click Search Now in the Error Inspector dialog box and note the number of errors reported that will need to be corrected. The Error Inspector will display a row for each feature that breaks a topology rule and identify which rule is being broken. Right-clicking a row will reveal many options such as zooming/panning to the error as well as ways to try to fix the error. By setting the Show option, you can list each type of error separately, and by selecting the Visible Extent only check box, you can zoom in on an area to isolate errors to correct.

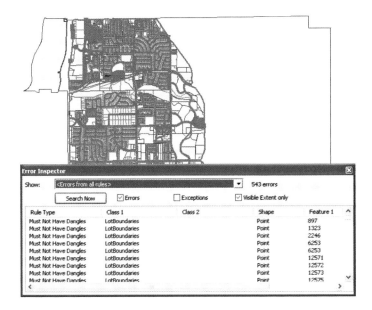

You will start with the Must Not Have Dangles errors and fix these in one small area at a time.

4 Load and zoom to the Topology1 bookmark where some errors have been identified. In the Show list, select LotBoundaries Must Not Have Dangles, and then click Search Now. A manageable 23 errors are shown. The Topology tools will give you options to correct multiple errors at once. You may need to extend a line to meet the boundary, trim a line that has gone past its boundary, or snap the end of a line to the boundary. There may also be circumstances where the lines are drawn correctly yet do not need to touch the boundary. These can be marked as an exception.

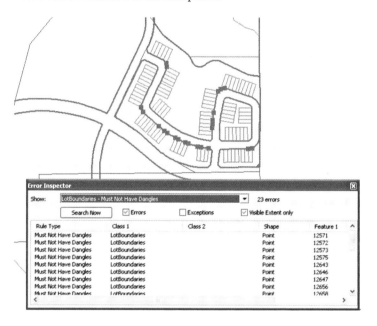

5 Select all the errors by clicking them in the Error Inspector dialog box. Multiple errors can be selected by pressing and holding the Ctrl key. Right-click, and then click Extend. Set Maximum Distance to **10** (feet) and press Enter.

6 Click Search Now again and note that all the errors have been corrected in this visible extent. In this case, ArcMap presented a suggested method for correcting the errors. This was to try to extend the lines until they intersected the closest linear feature. This is the same task performed by the Extend tool on the Advanced Editing toolbar. The difference here is that the topology rule allows this to be done in a batch process correcting all the lines at once. Note that ArcMap also suggested the Snap and Trim tools as possible fixes for these errors.

5-1

5-2

5-3

YOUR TURN

Clear the Visible Extent only check box. Using the same topology rule, click Search Now. Right-click a few of the errors and use Pan To to locate the errors on the map. Once again, select them all and try a mass correction using the Extend option.

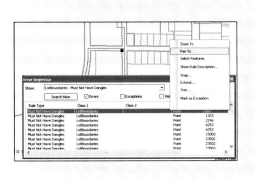

7 Click Search Now again. The remaining errors might be lines that extend past other lines rather than falling short. Select these errors and use the Trim tool with a distance of **10** feet to correct them.

8 The final errors may be like the one shown in the accompanying image. A small line segment extends from the other lot boundaries. This type of error can be deleted. Use Zoom To to locate the first one and verify that it is unnecessary. Right-click, and then click Select and delete it. The Extend and Trim options will take care of many of the errors. Always try these two options first before inspecting individual errors. There are still a few errors remaining, but these will be taken care of with a different method.

9 Load the Topology2 bookmark and zoom to it. Select the Visible Extent only check box and click Search Now. Click each row to highlight the errors in the map. These errors exist because the snapping distance was too small when the topology was created. Select the second error, right-click, and select Snap. Set the snap tolerance to **1** and press Enter.

10 Unlike the other fixes, this one requires that you revalidate the topology. On the Topology toolbar, click the Validate Topology in Current Extent button. This will limit the validation area and save time over having to validate the entire dataset. Confirm that the error is fixed by clicking Search Now.

YOUR TURN

Clear the Visible Extent only check box and do another search for **Must Not Have Dangles** errors. Zoom or pan to each error and use the tools you just learned to try and fix each one. You may need to use different distances for the Extend, Trim, and Snap tools. If any errors exist that cannot be fixed with these tools, mark them as exceptions. Validate the topology as necessary until all errors are corrected.

11 Zoom to Full Extent. In the Show list, select LotBoundaries Must Not Overlap, and then click Search Now. Load and zoom to the Topology3 bookmark, which will zoom to the first error on the list. Select the first error to highlight it on the map.

12 Right-click the error and select Subtract. This will give you an option to delete one of the overlapping segments. Click each segment and watch as it is highlighted on the map. Select the second feature and click OK. The overlapping segment of the feature you selected is removed, making a cleaner intersection.

5–1

5–2

5–3

YOUR TURN

All the **Must Not Overlap** errors can be fixed with the Subtract option. Work your way down the list, zooming in on and fixing each of the errors. When they are all fixed, repeat the process with **CityBoundary Must Not Overlap** and fix any errors you find. Finally, fix all the **LotBoundaries Must Not Self-Overlap** errors using the Simplify option, which is very similar to the Subtract option.

Note that the context menu for each type of error presented different options for correcting the errors, including an option to mark them as an exception. Exceptions will be maintained in the dataset as an error but will not be shown as an error in the Error Inspector.

Examine the polygon errors

The errors you have fixed so far have all been line errors. The remaining errors are polygon errors, and a different set of correction options is offered.

1 Load and zoom to the Topology4 bookmark. In the Error Inspector dialog box, in the Show list, select Parcels Must Not Have Gaps. Select the Visible Extent only check box, and then click Search Now. The size of your extents window may cause the Error Inspector to find more errors than are shown here. Click down the list and you'll notice that these errors appear as black polygons surrounding an entire block. These are created by the presence of streets. They are allowable gaps, so as you verify each one, right-click and set it as

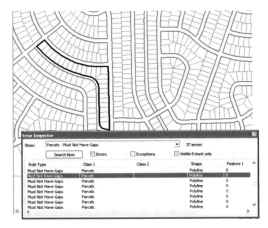

an exception. If you wanted to fix this error for the entire map, you could pan and search for Parcels Must Not Have Gaps errors and mark them as exceptions. This would take time because each one would need to be verified individually—a process that would eventually need to be done anyway in a production dataset. Once these errors are marked as exceptions, they will not come back in an error search. They will, however, still be maintained by the topology and could be found again by searching for exceptions.

2 Load and zoom to the Topology5 bookmark. Search for and clear all the Parcels Must Not Have Gaps errors in this extent, making sure to select the Visible Extent only check box.

3 In the Show list, select "Errors from all rules." Then click Search Now. There should now be fewer errors to tackle.

The errors include parcels, blocks, and subdivisions. Fix the parcels errors first, and then later you can fix the blocks and subdivisions.

4 In the Show list, select "LotBoundaries Must Be Covered by Boundary of Parcels," and then click Search Now. Turn off the Property_Data_Topology, Zoning Districts, and Subdivisions layers, and toggle the Lot Boundaries layer. You will see that there is a difference between the lot lines and the parcels. This occurred because the lot lines and parcels were created from the preliminary plat, and then the lot lines were altered later to follow the final plat. Once you've noted the changes, turn the Property_Data_ Topology and Lot Boundaries layers on

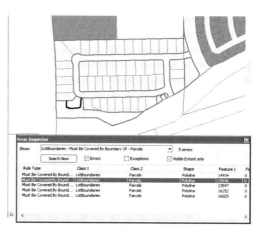

again. Now you will need to do the corrections. Select the error for feature number 23596. You will use this property line to split the parcels.

5-1

5-2

5-3

5 Right-click the selected error, and then click Select Features. Next, click the Split Polygons button on the Topology toolbar.

6 In the Split Polygons dialog box, click OK. Once again, validate the topology for the current extent. Notice that the parcels have been split at the property lines. That fixed the error where there was a property line, but no parcel edge. The other error is that there are two parcels that became a single parcel in the final plat. The solution here is to use the standard editing tools to merge the parcels together, and then once again validate the topology.

7 Set the selectable layer to Property Ownership and select the two polygons that will become the large park at the south of the subdivision as shown.

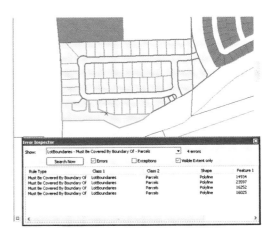

8 Now go to Editor > Merge and highlight the choices until the largest polygon flashes. Then click OK. This will merge the smaller polygon with the larger one, preserving the attributes. Validate the topology for the current extent, and you will see that the two parcels have become one. The map still displays many types of errors to correct, and you could pan through the dataset searching and correcting errors until all are eliminated, either by constructing the necessary features or marking them as exceptions. But for now, there is one more type of error to look at. The block boundaries for this subdivision are all producing an error. You'll use the Error Inspector to identify which topology rule is being broken, and then repair it.

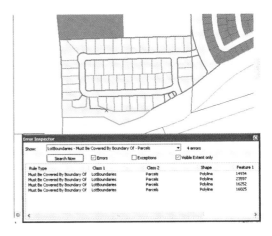

9 Use the Error Inspector to find errors for the rule Blocks Area Boundary Must Be Covered by Boundary of Parcels: Platted Property. Be sure to search the visible extent only. Highlight all the errors to see them on the map. The error exists for block polygons because they are covering parcels with a subtype of Plat Pending. The rule states that the block boundaries have to be over platted property. Once the parcel polygons are changed to the correct subtype, the error will go away. This is an example of a topology rule being enforced as a subtype.

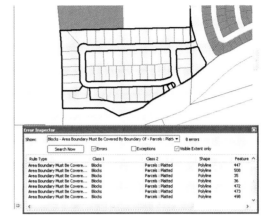

10 Right-click the selected errors and click Select Features. Then use the Select by Location tool to build the following query and select the parcels creating this error: "I want to select features from Property Ownership that intersect the features in Blocks that use the selected features."

11 Open the Attributes dialog box from the Editor toolbar. Change Plat Status to Platted Property for all the parcels. As a shortcut, click the Property Ownership title at the top of the list, and then change the Plat Status field. Accept the options to change the defaults to match the new subtype. This will change all the features together. Close the Attributes dialog box.

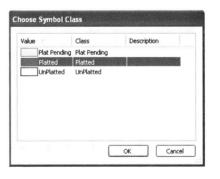

5–1

5–2

5–3

12 Validate the topology, and then search for the errors again. This should fix all but two of the errors, where the blocks do not follow the new parcel lines created in an earlier step. Highlight both errors, right-click, and then choose Select Features from the context menu.

13 The error can be fixed by merging these two polygons together. Use the Merge tool from the Editor drop-down menu and combine both pieces into block B. Then revalidate the topology and confirm that the errors are fixed. Close the Error Inspector.

Add new features

In tutorials 4–4 and 4–5, you created the lot boundaries for the Landing at Eden Lake subdivision. By importing those lines into the geodatabase, they can be used to create the new land parcels directly into the Parcels feature class.

1 Load and zoom to the Topology6 bookmark. Set the selection to Property Ownership and Lot Boundaries. Select the large polygon where the new parcels will go.

2 Use the Select by Location tool to select all the lot boundary lines associated with this parcel using the "share a line segment with" option. Click OK, and then delete the features, including the selected parcel.

3 Turn off all the layers except Lot Boundaries and make sure that you are in the data view. Add the lot lines for the new section to the map document. The file is called TheLanding and is in the My Answers folder in the City of Oleander geodatabase, inside the SampleData feature dataset. Right-click the new layer and go to Selections > Select All.

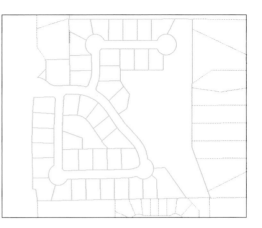

4 Click Copy, Edit from the main menu, and then click Paste. The template should be set as LotBoundaries by default. Click OK.

5 Turn off the The Landing layer and clear the selected features. Note that the lines are now part of the Lot Boundaries layer. Turn on the Property_Data_Topology and Property Ownership layers, and then validate the topology for the current extent. Many errors will be displayed for the new lot lines.

5–1

5–2

5–3

6 Open the Error Inspector. In the Show list, select "LotBoundaries Must Be Covered by Boundary of Parcels," and then click Search Now. Select all the reported errors, which will correspond to the new lines. The errors show that there are lot lines that exist without parcel polygons beneath them. The topology rules will be used to create the new polygons.

7 Right-click the list of errors, and
then click Select Features. Next, click
Construct Features on the Topology toolbar. Set Template to Platted and set
Cluster Tolerance to 0.09. Click OK. The
Construct Features tool created new polygons in an effort to fix the topology errors, in
much the same way that polygons were created to correct the map topology errors in
tutorial 5–1.

Now there are many errors where the new lot boundaries and parcels do not fit with the
existing lot boundaries and parcels. You will do some shared edits to correct this problem,
moving both the parcel boundaries and lot boundaries at the same time. By moving these
lines together, you will avoid creating any new topology errors and make the edits much
faster than if you only used the standard editing tools. You'll be moving the older lines to
match the new polygons you created.

8 Load and zoom to the Topology7 bookmark and run the validation process
again. You'll correct many of the errors
by moving the nodes into alignment
with the lines. Select the Topology Edit
tool. Move to the left area of the map display and while pressing N to restrict your
selection to topology nodes, click the
node shown here with a purple dot.

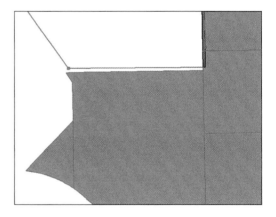

9 Drag the purple dot to the corner of the new
parcels, making sure to snap to the corner.
Validate the topology to see the results.

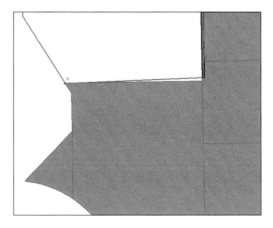

10 Move to the right corner and select the other topology node. Then move it to its correct position. Validate the topology and check your progress.

Using this process, you were able to correct an area where there was a gap between the parcels. The next process will correct areas where the parcels overlap.

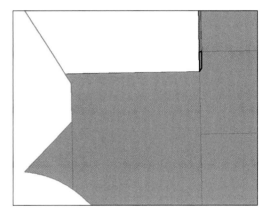

11 The red polygons show an error where the old parcels and new parcels overlap. Using the Fix Topology Error tool, select one of the red polygons.

To fix this error, you will merge the selected polygon with another polygon. At the same time, all the linear features that are coincident with the polygon boundaries will be adjusted. The Merge tool will let you highlight the adjacent polygons in order to determine which polygon should absorb the error.

5-1

5-2

5-3

12 Right-click and select Merge from the context menu. Click each of the polygons listed and watch as they highlight in the map. When you decide which polygon should absorb the error, highlight it and click OK.

YOUR TURN

Pan along the edge of the new parcels and select the next red polygon error. Correct this one using the Merge tool, and then correct the line work using the **LotBoundaries Must Not Have Dangles** rule as before. Validate the topology once you have corrected the error.

The last error was corrected using a combination of topology tools. Just as when you are constructing new features with the standard editing tools, it is important to understand the purpose of all the topology tools so that you can make the best selections for editing. Although there are still a lot of topology errors in this dataset, you've learned how to fix each of the different types using the topology tools. From this example, you can see that making a topologically correct dataset, and keeping it that way, is a lot of work.

13 Save your map document to **\GIST3\MyAnswers\Tutorial 5–3.mxd**. If you are not continuing to the next exercise, exit ArcMap.

Exercise 5–3

This tutorial showed how to create and validate a geodatabase topology. It also demonstrated many of the editing techniques to find and fix topology errors.

In this exercise, you will continue around the new subdivision and practice using the topology tools on some of the remaining errors. When you are comfortable with the tools, demonstrate each of the following procedures to the instructor:

- Demonstrate the use of the Error Inspector and restrict the results to one rule type.

- Mark an error as an exception.

- Select and move a topology node.

- Correct an overlapping polygon error.

- Correct a gap error.

- Demonstrate the Trim and Extend tools.

- Save your results as **\GIST3\MyAnswers\Exercise 5–3.mxd**.

WHAT TO TURN IN

There is nothing to turn in. The results of this lab are a visual demonstration of your new skills.

Tutorial 5–3 review

Of the three types of topology supported in ArcMap, the most sophisticated is **geodatabase topology**. This is constructed for existing feature classes and is stored in the geodatabase for future use. Unlike a map topology, a **geodatabase topology persists beyond the map document**.

A geodatabase topology can model behavior among points, lines, and polygons. There are also **rules for like features,** such as polygon-to-polygon behavior. ArcGIS contains a set of predefined rules to manage these relationships. You used some of the polygon and line feature topology rules to complete the exercise. There are a variety of rules available to customize how you want ArcMap to manage the relationships among features.

You first determined the relationships among the blocks, city area, city boundary, parcels, parcel boundaries, lot boundaries, and subdivisions. From these relationships, you determined which of the **topology rules** in ArcMap would accomplish what you needed. You used ArcCatalog to build the geodatabase topology. All the feature classes needed to exist in the same feature dataset, and feature classes can only participate in one topology.

When building topology, you **ranked** each of the data layers represented in the topology so that ArcMap knew which layers should be modified during the process. Typically, you rank the most accurate datasets higher. In this case, you selected Parcels and CityBoundary to have the highest rankings.

Once your topology was built, a new set of feature classes was created. These feature classes contain indicators of all the features that violate the rules you established, and the indicators were used to make corrections. By bringing the topology layer into ArcMap, and reviewing it in the **Error Inspector**, you were able to identify those features on the map that did not meet the topology rules you specified. You corrected these errors using tools provided by the Error Inspector.

The **Validate Topology** command was used to **ensure data integrity** by validating the features in a topology against the topology rules. When resolving errors, you have the option of marking an individual error or a collection of errors as **exceptions**. There will be instances when the occurrence of a defined error may actually be acceptable. In the exercise, the roadways created gaps among the Parcels polygons, which violated the **Must Not Have Gaps** rule. In this case, you marked it as an exception. Once an error is marked as an exception, it remains so until it is reset as an error. Running Validate Topology will not generate an error for an instance that has been marked as an exception.

STUDY QUESTIONS

1. While building topology, what would have happened had you given Subdivisions a higher ranking than CityBoundary?

2. List five geodatabase topology rules that were not used in the exercise and what feature type they pertain to.

3. Contrast and compare geodatabase topology with map topology and geometric networks.

Other real-world examples

Geodatabase topology benefits all organizations that rely on accurate spatial data. Since accuracy is paramount to most GIS applications, the ability to construct a clean fabric of features related to one another is critical to any successful GIS program. ArcMap and ArcCatalog have tools to ensure that data integrity is maintained among features within a geodatabase. Not only do topology tools ensure that you have the tools to build a sound dataset, but they also create an opportunity to affect the behavior among geographic features that participate in a geometric network.

River authorities, for example, use geodatabase topology to manage the relationships among streams, rivers, lakes, watershed boundaries, rainfall, and flow gauges. By setting up geometric networks between the edges and junctions of waterways, they can estimate the flow reaching reservoirs during and after heavy rainfall and mitigate flooding issues.

County appraisal districts use geodatabase topology to preserve taxing boundaries for developments within municipal boundaries that extend beyond the original county boundaries. New developments within a municipality that extend beyond the county boundaries are not subject to taxation by the original county. It is critical that the relationships between county and municipal boundary features are accurately portrayed and carefully managed to avoid inadvertently taxing these developments.

References for further study

There are a number of additional resources to help with topology. **ArcGIS Desktop Help** lets you use keywords to search for a title or topic. You can access ArcGIS Desktop Help in ArcMap or ArcCatalog by clicking Help on the main menu. Then click the Search tab, type a keyword, and click Ask. Use the keywords provided here to search for additional learning resources on many of the concepts taught in this chapter.

ESRI has developed a number of online courses on a wide variety of pertinent GIS topics. **ESRI Self-Study (Virtual Campus) courses** are an excellent resource for students that will supplement information covered in this course. ESRI Virtual Campus courses can be accessed at `http://training.esri.com/gateway/index.cfm`.

ArcGIS Desktop Help search keywords

topology, map topology, shared edits, geodatabase topology, spatial relationships, geometric networks, common topology tasks, editing a topology, topology in ArcGIS, topology validation, junctions, connectivity rules, network rules, logical network, utility network

ESRI Self-Study (Virtual Campus) courses

The following ESRI Virtual Campus classes may be helpful:

1. Working with Map Topology in ArcGIS
2. Working with Geodatabase Topology
3. Creating and Editing Geodatabase Topology with ArcGIS Desktop (for ArcEditor and ArcInfo)
4. Creating, Editing, and Managing Geodatabases for ArcGIS Desktop

Part 3
Optimizing the workflow

6

Customizing the interface

Tutorial 6–1

Customizing toolbars and menus

One of the simplest methods of customizing ArcGIS to fit your individual workflow is to customize the menus and toolbars that act as your interface to the software. Even though these are simple changes to make, they can have a dramatic impact on your ease in operating ArcGIS.

Learning objectives

- *Add new tools*
- *Modify the existing menus*
- *Create custom toolbars*

Introduction

There are several ways to modify the ArcGIS interface, with the simplest being to add more tools to the existing toolbars. There are many more tools available for ArcGIS than can be shown on the menus, and many powerful tools don't have a permanent home on the out-of-the-box default menus. If you look through the lists of available tools, you will no doubt find useful features that you will want to make available on a regular basis. This is as simple as dragging the newly discovered tool onto a toolbar.

Not all the tools have to be on the main toolbars, however. As you know, there are a host of context menus available throughout ArcGIS. These are the menus that appear when you right-click an item. And even though these menus are buried under other ones, they are readily accessible for customization.

Sometimes, the regular toolbar structure doesn't seem appropriate for the tools you may have selected, so ArcGIS allows users to create custom toolbars that will appear in the interface exactly like the default toolbars. You can even make submenus from the toolbars to stack a lot of functionality into a single custom toolbar.

Creating a new toolbar

Scenario   You've been getting questions from your staff about altering the ArcMap interface. It seems that the tools they use most are either not accessible or buried within context menus. This is making life difficult, and you want to find a way to ease their perceived burden.

Data   The map document for this tutorial has some inconsequential data to act as a backdrop while you work with the menus.

Tools used   ArcMap:

> Customize Mode
> Toolbars: Customize
> Toolbars: New Menu
> Toolbars: Context Menu

You will first take a look at the tools that are available inside ArcMap.

Research ArcMap tools

1   Start ArcMap and open the map document Tutorial 6–1.mxd. On the main menu, go to Customize > Customize Mode.

2   Click the Commands tab. The Customize dialog box has to be opened each time you want to change a toolbar. A quick way to open the dialog box is to double-click in any blank area on the toolbar.

The Categories pane shows groupings of tools. By selecting a category, the Commands pane will display a list of tools associated with those tools. Many of these tools do not appear on any toolbar, and will only be available if you add them.

6–1

6–2

3 Click a few of the categories and view the results in the Commands pane. If you find an interesting tool, click the Item Description link to get a brief synopsis of the tool's function.

The Toolbar Options button 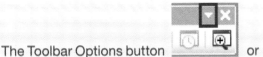 or, which is located on every toolbar, provides an easy shortcut to the Customize window. Simply click the button and select Customize to open the window. If the ArcMap toolbar's size prevents all the buttons from being displayed, you can still access them by clicking the Toolbar Options button to make them visible.

4 Scroll down and find the category Pan/Zoom. In the Commands pane, drag the Continuous Zoom and Pan tool to the main menu next to the standard Zoom and Pan tools. A thick bar will appear to show where the tool's icon will display. A circle with a slash through it indicates that the tool cannot live there. When you have the tool in an appropriate spot, release the mouse. Close the Customize Mode dialog box and experiment with the new tool. This tool was easy to find, but some may be difficult to find, or may be unknown to you. In the Customize Mode dialog box, there is a Search tool to help locate tools.

5 Open the Customize Mode dialog box again, click the Commands tab, and type **Select Features** on the line labeled Show commands containing. A list of categories containing tools with those words in their name is displayed in the Commands pane. Choose the Selection category, and then drag the Select Features on Screen tool from the Commands pane to the Tools toolbar next to the Clear Selected Features button.

6 Next, right-click the new tool and select Begin a Group. This will place a bar between the tools and make them more visible. Close the Customize Mode dialog box and test the tool.

The new tool fits right in on this toolbar. Another concern of the staff was not to have their favorite tools spread all over the interface. You'll fix this by creating a custom toolbar and adding their favorite tools to it.

7 Open the Customize Mode dialog box and click the Toolbars tab. A list of all the toolbars available from the default ArcMap installation is displayed. On the right side, click New and name the new toolbar **Oleander**. The new toolbar will be docked at the top of the map area.

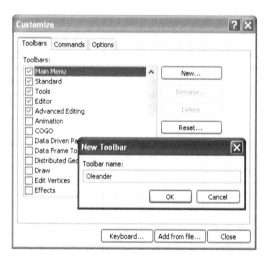

6–1

6–2

8 As long as the Customize Mode dialog box is open, you can add tools to the new toolbar. Undock it and drag it into the map area so that it is easily visible. Click the Commands tab, and in the Categories pane, click Editor. Then drag the Save Edits tool from the Commands pane to your new toolbar. When the tool is placed on the new toolbar, right-click it and select Image and Text.

9 Add Start Editing and Stop Editing to the toolbar as well, with both the image and text displayed. Note that it is possible to rename a tool on your custom toolbar. Close the Customize Mode dialog box.

Your new toolbar will function just like a standard toolbar. It can be moved around, docked, closed, and added from the Toolbars menu under the View tab.

YOUR TURN

Add a tool to the Oleander toolbar to turn the snapping on and off. (**Hint:** Search for snapping commands and view their descriptions). Then add a tool next to it to set the selectable layers.

Add a drop-down menu

A drop-down menu can be added to either the standard menu or a custom menu. For the benefit of your staff, you'll add one to the Oleander toolbar that switches the map display from the Data View to the Map View.

1 Open the Customize Mode dialog box and click the Commands tab. Scroll all the way to the bottom and click New Menu. In the Commands pane, click New Menu and drag it to the Oleander toolbar, releasing it at the bottom.

2 Right-click the menu and change the name to **Select View**.

3 Scroll up the Categories list and click View. Drag the Data View tool up to the Select View menu and a new box will open below it. Drop the tool there. Next add the Layout View tool to the menu. Close the Customize Mode dialog box and try out your new drop-down menu.

Customize a context menu

The final modification you'll make to the menus and toolbars will be to add a tool to the context menu for layers.

1 First, investigate where the new tool will go. Right-click any layer and look at the context menu. You will add a tool below Open AttributeTable that will open the table and display only the selected features.

2 Open the Customize Mode dialog box and click the Toolbars tab. Scroll down the list and select the Context Menus check box. A new toolbar will appear. Drag it into the map area and click the drop-down arrow. Scroll down the list and find Feature Layer Context Menu. This is the context menu you investigated earlier. Click it to see the tools it contains, and then click the Copy tool. This will keep the context menu open while you add tools to it.

6–1

6–2

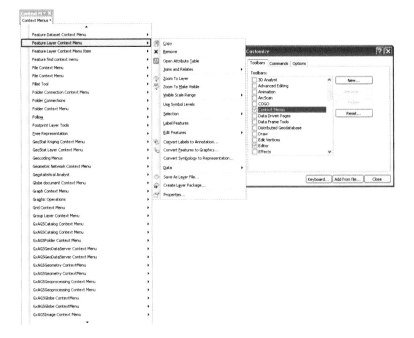

3 In the Customize Mode dialog box, click the Commands tab, and then click the Layer category. Drag the Open Table Showing Selected Features tool from the Commands pane to the context menu. If the context menu closed, you'll need to hold down the left mouse button and retrace your steps to get the context menu open again. Close the Customize Mode dialog box.

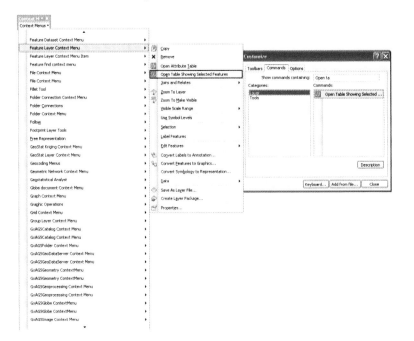

4   Click the Select Features on Screen tool you added earlier, right-click the Lot Boundaries layer, and then try out the new tool. These new menus and tools will be of great help to your GIS staff.

5   Save your map document as **GIST3\MyAnswers\Tutorial 6–1.mxd**. If you are not continuing to the next exercise, exit ArcMap.

Exercise 6–1

The tutorial showed how to create a new menu in the ArcMap interface and add tools to it.

In this exercise, you will repeat the process to create a custom editor menu. Many of the editing tools that you might want to use on a single project are loaded onto different menus. To speed up your editing, you'll create a custom menu and add your own set of tools to it.

- Start ArcMap and open Tutorial 6–1.mxd.

- Create a new menu called **Fast Edits**

- Add these tools to your new menu from the Customize Mode dialog box:

  ▪ Save Edits

  ▪ Edit tool

  ▪ Rotate tool

  ▪ Curve Calculator

- Save your results as **\GIST3\MyAnswers\Exercise 6–1.mxd**.

6–1

6–2

WHAT TO TURN IN

If you are working in a classroom setting with an instructor, you may be required to submit a screen capture showing the menu you created in tutorial 6–1.

A screen capture of the menus you created in

Tutorial 6–1

Exercise 6–1

Tutorial 6–1 review

When attempting to accomplish a task in ArcMap, you may need functionality that you cannot find in the standard sets of menus and toolbars. This does not mean that a tool does not exist—it simply is not accessible from the standard interface. There are hundreds of tools available in ArcMap to manage spatial features, and if you were to show every tool, there would not be enough room in your map document to work with your data!

ArcMap groups tools in categories according to their primary functions. They are found by going to Customize > Customize Mode and clicking the Commands tab. There are many beneficial tools there that do not appear on the regular menus.

One of the simplest methods for customizing ArcGIS is to customize the menus and toolbars by adding more tools to the existing toolbars. You accomplished this simply by dragging a needed tool from the Customize Mode dialog box to the toolbar.

You created your own **customized toolbar** (Oleander) by clicking the Toolbars tab under Customize > Customize Mode and clicking the New button. You then created a **customized drop-down menu** for that toolbar that allowed you to switch between Map View and Layout View.

You created a **custom context menu** (these are the menus that appear when you right-click an item) in a similar fashion by first creating the context menu using the Customize Mode dialog box and selecting the Context Menus check box under the Toolbars tab. Once finding the appropriate tool, you just dragged it onto the context menu.

STUDY QUESTIONS

1. List four tools to place on a customized toolbar that would help you become more efficient using ArcMap. What is the objective of this new toolbar?

2. If you had a chance to redesign the ArcMap interface to include different tools available on the standard toolbars, how would you do it? Would you choose to display fewer tools or more of them?

3. Do you think you could streamline the interface for non-GIS professionals and make their job easier? Or would you simply teach others where to find the tools they need? Explain your answer.

Other real-world examples

Creating custom menus and toolbars is a great way to save time and make you more productive. In fact, by creating a number of customized tools (which you will learn about in the next tutorial) and placing them on one (or several) toolbars, you can begin to develop customized extensions for ArcMap.

Organizations will often customize the ArcMap interface to tailor the interface and reduce its complexity for employees who perform specialized tasks. A municipality with a group of users dedicated to editing the sanitary and storm drain systems, for example, will customize the ArcMap interface to present only the necessary functionality for these employees to complete their tasks.

to achieve a common goal operating within the ArcGIS framework to extend the functionality of the standard ArcGIS software. A variety of vendors provide suites of tools that extend the functionality of ArcGIS into niche, or specialized, areas. ESRI provides a wide variety of extensions that work with ArcMap, including Spatial Analyst, 3D Analyst, Survey Analyst, Maplex (see tutorial 8–1), Geostatistical Analyst, Network Analyst, and many more. Over the years, many of the more popular tools that were once confined to extensions are now contained within core ArcGIS products.

6–1

6–2

Tutorial 6-2

Creating toolbar scripts

With custom scripting, you are not limited to the tools provided in ArcGIS. Scripting gives you the ability to make custom tools to place on your custom menus. There are also ways to make the custom tools function just like the default tools.

Learning objectives

- Add custom scripts to toolbars
- Configure custom tools
- Run scripts from a button
- Import scripts

Introduction

Even though there are dozens of tools provided in ArcGIS, you will inevitably come across a process that is not done by the stock tools. By using scripting, either through Python or another scripting language, you can create custom tools and place them on a menu. You can also find scripts authored by other ArcGIS users that you could incorporate into your own project.

This tutorial will focus on writing a simple Python script using the IDLE interface. ArcGIS Desktop Help contains a description of all the processes available to use and provides a scripting example written in Python. Often, your custom script may only be a few of the tool examples strung together into a new tool.

The Python interface in ArcGIS is called ArcPy. This is a module for Python that adds all the geoprocessing tools of ArcGIS into the Python environment. The first step in any custom tool script is to import this module into Python with the command "import arcpy." After this is added, any of the ArcGIS tools, as well as native Python functions, can be invoked in the script.

Custom tool scripts are stored in a toolbox created by the user. Once a script is added to the toolbox, it can be configured to accept user input. Any parameter of an ArcGIS process can be made user-definable, giving scripts great flexibility. Scripts can also include flow of control statements, such as if–then or "do while" statements, to control how the process will be completed.

Once a script becomes a tool in a toolbox, it is a very simple process to add it to any toolbar. While it is not always convenient for users to run tools from the toolbox, having a toolbar button in ArcMap will make the tools easily accessible.

Making a script into a tool

Scenario   The staff was very impressed with the lesson on adding stock tools to the menus, and they've all gotten pretty good at it. In an effort to remain the top dog, you will need to master the following tricks to add scripts to the menus and buttons.

The Oleander Police Department staff does a lot of statistical analysis of their crime data. One step in the process is to create a 528-foot (tenth of a mile) grid for the City of Oleander. Then they run other processes and compile statistics for each grid cell. When they analyze the next set of data, they need to create another grid.

Creating a grid each time is getting tiresome, and they would like a script that will do this for them. It needs to cover the main part of Oleander, which is about 2½ miles east to west and 3½ miles north to south. You'll make the grid name user-definable so that they can run this grid model many times and not worry about overwriting previous analyses.

Data   This tutorial contains background data for interest, but it has no effect on the functionality of the tools.

Tools used   ArcMap:

> Toolbars: New Menu
> Script Editor
> Toolbars: Geoprocessing tools
> Toolbars: ToolTips

Create a custom tool

The IDLE interface is used to write Python scripts. Once the script is written, you will add it to a custom toolbox.

1   From the Start menu on your computer, browse to ArcGIS 10 > Python 2.6 and click IDLE. This will open the Python programming interface.

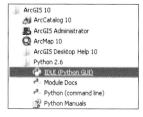

The IDLE interface will let you type lines of code to run and test. Since you will want to save the code to a file for later inclusion in an ArcGIS toolbox, you will need to add a scripting window that will let you save your results.

2 In the IDLE window, click File > New Window.

In this window, you will create and save the script for the Oleander Police Department. First, you will add some information about the origin of the script, and then you will load the ArcPy module.

3 Add the following lines at the top of the script. Lines prefaced by a pound sign (#) are notes and do not run.

- `# Name: CreateFishnet.py`
- `# Description: Creates a temporary grid for the Police Department`
- `# Description: The grid is a one-tenth mile grid`
- `# Author: -your name>`

4 Next, add the line that loads the geoprocessing module ArcPy:

- `# Import system module`
- `import arcpy`
- `from arcpy import env`

Next, you will want to specify where to store the output file that is created. By setting the workspace environment, the script will automatically place any output files there, as well as look there for any other referenced file.

5 Add the lines to set the workspace location:

- `# Set workspace environment`
- `env.workspace = "C:/ESRIPress/GIST3/MyAnswers /CityOfOleander.mdb"`

(**Note:** Change this as necessary to match your install location.)

At this point, you need to know the syntax of the Create Fishnet tool. This will show you what parameters the tool needs to run, and in what order. You will need to make variables to handle all the parameters.

The syntax can be found in the tool's Help file, but for convenience, it is provided below. The line you add to the code will be formatted exactly the same as the syntax that is shown.

```
Syntax
CreateFishnet_management (out_feature_class, origin_coord, y_axis_coord, cell_width, cell_height,
number_rows, number_columns, {corner_coord}, {LABELS | NO_LABELS}, {template}, {POLYLINE |
POLYGON})
```

Now, you will make a variable for each required parameter. To define a variable in Python, you simply type the name and equal sign and what value the variable should hold. You do not need to define a variable type or use any other keywords.

6 Type these variable names and values. The values were provided by the police department based on what it has used in the past.

- outFeatureClass = 'PoliceGrid'
- originCoordinate = '2397370 6989659'
- yAxisCoordinate = '2397370 6989669'
- cellSizeWidth = '528'
- cellSizeHeight = '528'
- numRows = '35'
- numColumns = '25'
- oppositeCorner = '#'
- labels = 'true'
- templateExtent = 'City _ Limit'
- geometryType = 'POLYGON'

(**Note:** The pound sign in this case means that no value is being provided. This is acceptable in parameters that are optional.)

6–1

6–2

There is a problem with the output feature class name as provided by the police. This would have the script writing the same file name every time, but you would like the file name to be provided by the user. To do this, you will set the variable equal to a command that will get the value from the user.

7 Change 'Police Grid' to **arcpy.GetParameterAsText (0)**.

Finally, add the Create Fishnet command. As you get to the end of a line, press Enter to add more parameters on the next line. The script will continue to read the parameters until it finds the closing parenthesis to mark the end of the command.

8 Add the Create Fishnet command as shown:

- ```
  arcpy.CreateFishnet _ management(outFeatureClass,
  originCoordinate, yAxisCoordinate, cellSizeWidth,
  cellSizeHeight, numRows, numColumns, oppositeCorner,
  labels, templateExtent, geometryType)
  ```

  ```

9 Save the file in your Scripts folder as **CreateFishnet.py**. Close both Python windows. With the script complete, you will need to add it to a toolbox for it to be usable inside ArcMap.

10 Start ArcMap and open the map document Tutorial 6–2.mxd. Open the Catalog window and pin it in place by clicking the Auto Hide button.

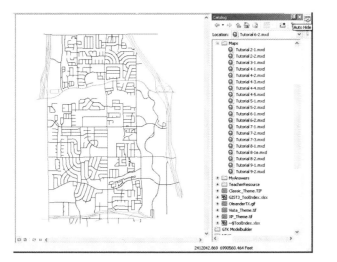

11 Scroll down and expand the Toolboxes folder. Right-click MyToolboxes and select New to create a new toolbox. Name it **Police Tools**.

12 Right-click the Police Tools toolbox and select Add > Script.

6–1

6–2

13 Name the script **PoliceGrid**, with a label of **Police Grid**. Add a description of **Creates a 528-foot grid for crime analysis**. Then select the check box to store relative path names. Click Next.

14 Use the Script File input to browse to your new script file. **Click Next**. This last window is used to set up the user-definable variables. The value retrieved by the GetParameterAsText command will be stored in the location defined here.

15 Click the first row under Display Name and type **NewFileName**. Press Tab and set Data Type to Any value. Notice in the Parameter Properties pane that this value is required and set as input. **Click Finish**. The new script appears in the Police Tools toolbox. Double-clicking the script will run it, just like any other tool in ArcMap. Right-clicking the script and selecting Edit will open the IDLE interface, in case changes need to be made.

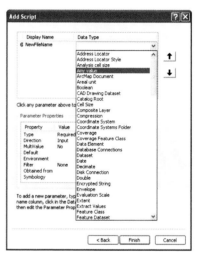

16 Double-click the Police Grid script. Enter a new file name, such as **PoliceGrid01**. The script should run successfully and add the new layer to the map document. If you experience any errors, edit the script from the toolbox and check it against the commands above.

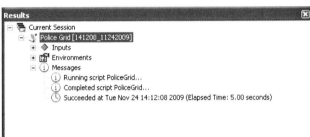

You succeeded in making a script that will automate one of the police department's most common functions. But it is tedious to have to locate this tool in the toolbox every time staff would like to run it. Instead, you will make it into a button on an ArcMap toolbar.

17 Open the Customize Mode dialog box. Go to the Commands tab, scroll to the bottom, and select Geoprocessing Tools.

6-1

6-2

18 Click Add Tools and navigate to the Police Tools toolbox. Select the Police Grid tool and click Open. The script is now added as an ArcMap geoprocessing tool.

19 Add it to the Tools toolbar by dragging it and dropping it next to the Full Extent button.

20 Right-click the tool and change the name to
Create Police Grid. Change the display
setting to Image and Text, and change
the button image. Click Close to close the
Customize Mode dialog box. Clicking the
button will run the script you wrote. This
will be a lot more convenient for police
staff to use. Note that pausing over the tool
produces a ToolTip.

Enhance the script

The script works well, but now the police staff are asking that you make the cell size user-definable. Sometimes, they use a 660-foot grid, and sometimes they use a 1-mile grid, depending on whether they are using state or national data. You'll modify the script to allow them to change the cell size.

Scripts can be edited in ArcMap through the Catalog window. From there, you'll be able to open the script, make the necessary changes, and save and test it.

1 In the Catalog window, right-click the Police Grid script and select Edit.

2 You can close the Python Shell window, since all the edits will be done in the other Python window.

3 Move the cursor just below the lines that set the environment workspace and add these lines:

- # Define a variable to accept new user input
- cellsize = arcpy.GetParameterAsText (1)

```
# set workspace environment
env.workspace = "C:/ESRIPress/GIST3/MyAnswers/CityOfOleander.mdb"

# Define a variable to accept new user input
cellsize = arcpy.GetParameterAsText(1)
```

Note that the GetParameterAsText method has been incremented to 1. This is to distinguish it from the other user input line. If more user inputs were added, they would increment to 2, 3, and so forth.

4 Now change the values of CellSizeWidth and CellSizeHeight to this new variable. Then save and close the script. That was pretty simple. Note that you could have replaced the variable names in the command line for the CreateFishnet command instead of setting the existing variable to the new variable. Both methods achieve the same result.

```
# set workspace environment
env.workspace = "C:/ESRIPress/GIST3/MyAnswers/CityOfOleander.mdb"

# Define a variable to accept new user input
cellsize = arcpy.GetParameterAsText(1)

outFeatureClass = arcpy.GetParameterAsText(0)
originCoordinate = '2397370 6989659'
yAxisCordinate = '2397370 6989669'
cellSizeWidth = cellsize
cellSizeHeight = cellsize
```

6–1

6–2

Next, you will need to modify the input parameters of the script to accept the new user input.

5 Right-click the script again and select Properties. Go to the Parameters tab.

6 In the Display Name column, add a new entry called **CellSize**. Set Data Type to **Any value.** This adds a new entry to accept user input. Note that these need to be in the same order as the increment number in the script. In this script, increment 0 accepted the file name and increment 1 accepted the cell size. If more inputs are added to the script, the parameters list must maintain the same order.

Now, the script is ready to run.

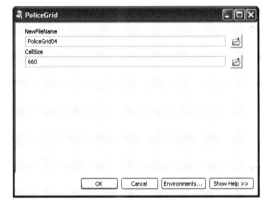

7 Run the script from the button you created earlier. Note the additional input fields, as well as the order. Enter a cell size of **660** and a file name of **PoliceGrid04**.

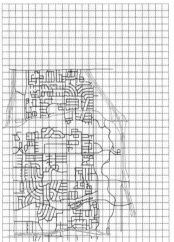

8 Add the new grid to your map document to see the results of the script. The new grid now has a larger cell size to accommodate whatever the police staff's needs are.

YOUR TURN

The police staff are happy with the script, but with the larger cell size, they don't need as many rows and columns. Making the number of rows and columns into user-definable parameters will solve the problem.

Add two more variables to the script to accept the values for the number of rows and the number of columns. Then add these to the script parameters to accept the inputs. Make sure to get them in the correct order.

This tutorial has shown how to write a Python script and include it as a tool in a custom toolbox. It also showed how to make that tool a button on a toolbar with a custom name and icon. You can try making other geoprocessing scripts by reviewing the scripting examples of other geoprocessing tools and combining them with user-defined parameters.

9 Save your map document as **\GIST3\MyAnswers\Tutorial 6–2.mxd**. If you are not continuing to the next exercise, exit ArcMap.

6–1

6–2

Exercise 6–2

This tutorial showed how to write a Python script to perform a task and how to make the parameters of the tool's process user-definable. It also demonstrated how to place the script on a toolbar for convenience.

Now, the police staff have another request. It seems that in the course of their data analysis, they select a single grid cell and perform statistical reviews. Then they select all the street centerlines that intersect that cell and perform more statistical reviews. They can easily select the single grid cell but would like a button on the toolbar that will select all the intersecting street centerlines for them.

The script will use the Select by Location tool. It will also require user input to identify which grid file is being used and which process to select all the streets that intersect the selected cell.

- Review the command syntax for the Select by Location tool. Note that you must convert the input feature class to a layer file within the script.

- Create a new script that will perform the process. (**Hint:** Copy the code sample in the Help file, and then modify it for this application.)

- Add the script as a tool to the Tools toolbar.

Be able to demonstrate the tool and show the underlying code.

- Save your results as **\GIST3\MyAnswers\Exercise 6–2.mxd**.

WHAT TO TURN IN

If you are working in a classroom setting with an instructor, you may be required to submit the work you created in tutorial 6–2.

Demonstrate to the instructor the toolbars created in

Tutorial 6–2

Demonstrate the script created in

Exercise 6–2

Tutorial 6–2 review

In the first exercise, you learned how to customize the ArcMap interface by creating **custom toolbars** and **customized context menus**. Being able to customize the interface enables you to access tools that more closely resemble your workflow. This exercise focused on going a step further and building upon your knowledge of **creating toolbars** to **create custom tools** (buttons) that can reside within those toolbars.

While hundreds of useful tools are available, the key to success is learning how to incorporate the tools you need into your ArcMap document so that you can use them. Adding custom functionality to buttons and toolbars is a straightforward and standard operation.

Writing Python scripts is probably the best way to customize your ArcGIS processes. These scripts are not difficult to write, and a host of good examples can be found in ArcGIS Desktop Help and on the Internet. The scripts can accept user input, produce output, and as you'll see in tutorial 7–2, they can control the flow of the processes.

Congratulations. You are well on your way to mastering one of the best ways to stretch the efficiency and functionality of ArcGIS.

STUDY QUESTIONS

1. There are a variety of Web sites available that contain scripts that can be imported into ArcGIS. List an example and describe the process for downloading a script from this Web site.
2. What other programming languages does ArcGIS support?
3. How would you make scripts available in other map documents?

6–1

6–2

Other real-world examples

The basic components of ArcGIS Desktop, including ArcView, ArcEditor, and ArcInfo, contain most of the tools you'll need to perform complex analyses. These components can be extended through custom-built tools. There are hundreds of tools written by ArcGIS users around the world that supplement the standard functionality of ArcGIS. Many of these tools can be downloaded for free at `http://arcscripts.esri.com/`. The availability of aftermarket products that users build and make available to other users is one of the most v aluable aspects of ArcGIS. Chances are good that if you are in need of a custom tool to accomplish your task, someone somewhere has already built a tool to do it.

A pipeline company needed to develop a tool that would allow non-GIS staff to input pipeline replacement data into an SDE geodatabase. A custom toolbar, containing several tools for manipulating pipeline data, guides staff through all the necessary steps to replace a segment of pipeline in the geodatabase in an accurate and standardized fashion.

A public agency that heavily uses Global Positioning System (GPS) technology has developed an extension for ArcMap that incorporates several customized tools for input, management, and export of GPS-derived data. Written with standard programming languages, this extension to ArcMap is widely used by other public and private organizations around the country.

References for further study

There are a number of additional resources to customize the ArcMap interface. **ArcGIS Desktop Help** lets you use keywords to search for a title or topic. You can access ArcGIS Desktop Help in ArcMap or ArcCatalog by clicking Help on the main menu. Then click the Search tab, type a keyword, and click Ask. Use the keywords provided here to search for additional learning resources on many of the concepts taught in this chapter.

ESRI has developed a number of online courses on a wide variety of pertinent GIS topics. **ESRI Self-Study (Virtual Campus) courses** are an excellent resource for students that will supplement information covered in this course. ESRI Virtual Campus courses can be accessed at `http://training.esri.com/gateway/index.cfm`.

ArcGIS Desktop Help search keywords

customizing the interface, custom toolbar, context menus, toolbox, scripts, ToolTips, MapTips, custom dialog boxes, customizing, new menu toolbar, adding tools, programming tools, Python, programming languages, macros

ESRI Self-Study (Virtual Campus) courses

The following ESRI Virtual Campus classes may be helpful:

1. Customizing ArcGIS Desktop
2. Customizing ArcMap: Easy Ways to Extend the Interface
3. Getting Started with Scripting in ArcGIS 9

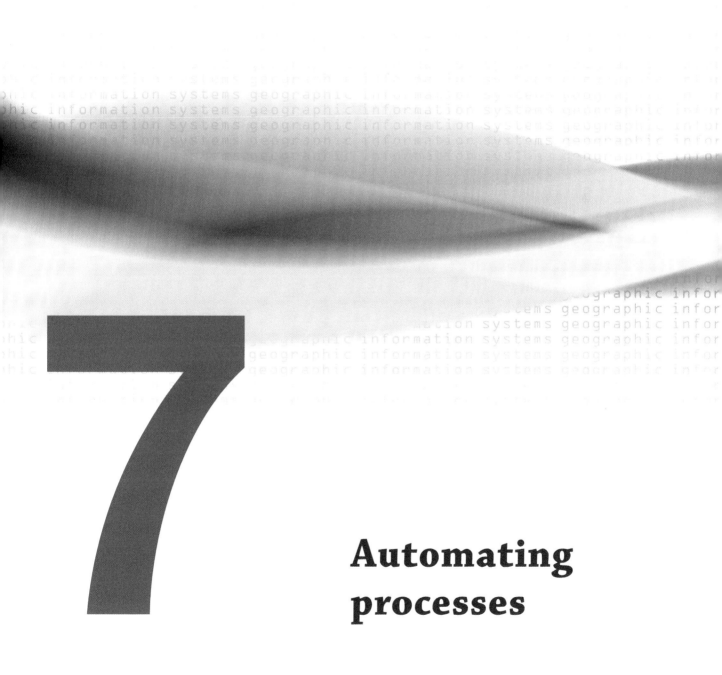

7

Automating
processes

Tutorial 7–1

Getting started with ModelBuilder

ModelBuilder is a programming tool that can be used in ArcGIS to develop a custom task or automate a workflow. Models are built through a visual interface, which makes them easy to understand and develop. Models can incorporate any of the geoprocessing tools as well as custom scripts written by users.

Learning objectives

- *Learn concepts of ModelBuilder*
- *Assemble ModelBuilder components*
- *Work with variables*
- *Create custom tools*

Introduction

An important thing to understand about ModelBuilder is that it does not add any new functionality to ArcGIS. Rather, it lets you build a model to combine existing functionality to simplify overall use. When you create a model, you should first go through the steps manually to make sure that the desired process can be accomplished with ArcGIS tools. Only then will you know what steps are involved in the process and which tools need to be included in the model.

When tools are added to a model, they require variables to act as the input and output data. By using the output variable from one tool as the input variable for another, a complex string of tools can be used to perform several processes. Additionally, these tools can be made into parameters of the model and be defined by the user. When the model is run, the user will be prompted for values, which may include the name of the input or output feature classes or other parameters required by the tools in the model.

As you build a model, you will notice that it can exist in one of three states. The first state, indicated by the lack of color in the model components, is the Not Ready to Run state. The tools may lack one or more required parameters and therefore would not run if asked to do so. Once all the required parameters are set, the model moves into the Ready to Run state, indicated by the addition of color to the model components. After the model has successfully run, the components take on a gray drop shadow indicating that the model is in the Has Been Run state.

Once your model is complete and its processes have been verified, it can be run from the model window in ModelBuilder or from the Search or Catalog windows in ArcMap. Models can also be written to perform tasks in ArcCatalog. You've also learned how to run your model from an ArcMap toolbar (see tutorial 6.2). It is important to note that once you include model parameters to prompt the user for input, you should no longer run the model from the model window. Doing so will not generate an input screen, and the model will be run using only the default setting.

Creating a new toolbox

Scenario   The City of Oleander's library has a bookmobile that it takes around town to apartment complexes to serve that segment of the community. Before the bookmobile is scheduled for a stop at a particular complex, the library mails out a notification to all the regular library patrons within a half mile and invites them to take advantage of the visit. Some patrons order books online and take delivery through the bookmobile when it is in their neighborhood, and others save a trip downtown by returning books there.

The library has a geocoded dataset of all active library patrons, and the GIS department has provided the locations of all the apartment complexes. The library would like you to develop a process to help with patron notification. Since the bookmobile has a scheduled stop once a week, the notification is done many times a year, and it would be beneficial to automate the process with a model.

First, the librarian will select the apartment complex for the next scheduled bookmobile appearance. Then it is buffered by 2,640 feet, and the resulting buffer is used to select the patrons inside the buffer. To wrap things up, a table is written containing the addresses of the selected patrons for use in a mail-merge document.

Data   The library staff maintains a point shapefile of the patron locations, which is updated every month. The data is aggregated to the address field along with a summary of how many times each patron visited the library that month, stored in the Number of Visits field.

The GIS staff maintains a polygon feature class of the apartment complexes.

Other data such as the street centerline file and parcel data is provided for background interest.

Tools used   ArcMap:

New Toolbox
ModelBuilder tool

7-1

7-2

7-3

When you create a new model, it is stored in a toolbox. Users must create their own toolbox for custom models and scripts, so you'll do that first.

Create a user toolbox

1 Start ArcMap and open Tutorial 7–1.mxd.

2 Click the Catalog tab at the right side of the map window. Navigate through the catalog tree and find Toolboxes. Expand it to show My Toolboxes.

3 Right-click My Toolboxes, and then click New Toolbox. Name the toolbox Library Models. It will automatically get a .tbx extension.

4 The Catalog window will automatically retract when you move to work in another area of ArcMap. Since you'll be doing a lot of work in the window, pin it in place by clicking the Auto Hide pin. The catalog tree will remain visible until you unpin it.

The library manager has already verified the process and selected the tools that you will need. Here are the steps, along with the required tools:

Step	Tool
Select an apartment complex	User Selection
Buffer the selected polygon	Buffer
Select patrons within the buffer	Select Layer by Location
Create an output table	Copy Rows

Notice that some of the geoprocessing tools do not carry the same name as their ArcMap menu counterparts. For instance, Select by Location from the ArcMap menu becomes Select Layer by Location as a model tool.

Set up the environment

First, a little housekeeping to make the model building experience go a little more smoothly. The first setting will help to identify which parameter to use when tools are connected in the model window. The second will set up the default workspaces to store the results.

1 On the main menu, go to Geoprocessing > Geoprocessing Options. Select the "When connecting elements ..." check box and the "Overwrite the outputs of geoprocessing operations" check box. Keep the pre-existing check marks. Click OK to close the dialog box. Notice the selection for Background Processing. When enabled, this will put geoprocessing tasks, such as models, into a background process and free up ArcMap so that you can continue working. The slider bar controls how long the process window is displayed before it is automatically minimized.

2 On the main menu, go to Geoprocessing > Environments. Click Workspace and set the Current Workspace and Scratch Workspace to the CityOfOleander geodatabase in the MyAnswers folder. Click OK to save the environment settings and close the dialog box.

Now it's time to start the model creation process.

7-1

7-2

7-3

Build a new model

1 Right-click the Library Models toolbox and go to New > Model.

2 In the model window, go to Model > Model Properties. Type the name of the model as **PatronNotification.** Type the label as **Patron Notification**, and type a description as follows:

> **The user selects an apartment complex and this model generates a list of library patrons within 1/2 mile.**

3 Select the check box to store relative path names and click OK. Save the model by clicking **Model > Save**. Notice that when you saved the model, the name of the editor window changed, as well as the name displayed in the Catalog window. Always save your model, and always give it a descriptive name. If you wind up with twenty models called Model 1 through Model 20, you will have a hard time later finding the one you need for a particular job.

Add tools to the model

1 Using the Search window, find the Buffer tool. Drag the Buffer tool into the model. It will appear without color, meaning that it is in the Not Ready to Run state.

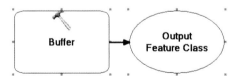

2 Double-click the Buffer tool to open it and set the parameters. In the Input Features list, click ApartmentComplexes, and set Distance to **2640** feet. Accept the defaults for the other settings (including Output Feature Class) by clicking OK.

The components of the model will now appear with a color fill. This indicates that they are in the Ready to Run state, but there are more tools to add to the model.

3 Locate the Select Layer by Location tool and drag it into the model window just below the Buffer tool. You may need to zoom out a bit. Then click the Connect button.

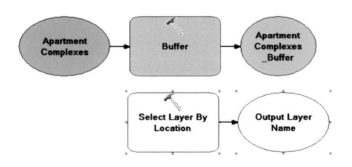

4 With the Connect tool, click in the Buffer output variable, and then in the Select Layer by Location box. In the resulting box, click Selecting Features to identify where the output buffer will be used. Click OK.

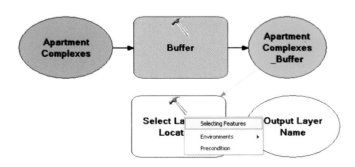

7–1

7–2

7–3

5 The new components of the model are still in the Not Ready to Run state, because there are still some required parameters that have not been set for the selection. Switch to the Select tool and pause the cursor over the Select Layer by Location tool to see a report of its status. Notice that there are several parameters that are not set, such as the input and output layer names.

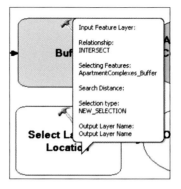

6 Double-click the Select Layer by Location tool to open it. Notice that Selecting Features has already been set with the Connect tool. Click the Input Feature Layer drop-down arrow and select Aggregated Patron Locations. Take a moment to note the other options for Relationship and Selection type. These mimic the selection settings from the standard geoprocessing tool. Click OK.

7 The model may be a little confusing or run out of the window. On the Model toolbar, click the Auto Layout button, and then the Full Extent button to make the display a little clearer.

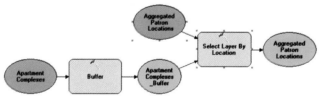

8 The final step in the model will be to write the results to a new table. Find the Copy Rows tool and drag it into the right side of the model. Then use the Connect tool to connect the selection output variable to the new tool. Set the input parameter as Input Rows.

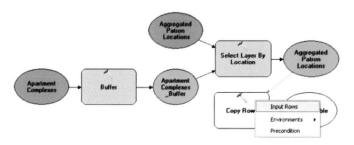

9 Next you need to define a setting for the outputs to determine whether they will be saved permanently and if they need to be displayed in the model. Right-click ApartmentComplexes_Buffer. Make sure Intermediate is selected and click **Add to Display.** Data set as Intermediate will be saved as a feature layer file, mean-

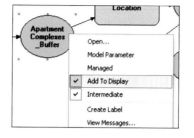

ing that it persists for the life of the MXD file, but it is not permanently saved to the hard drive. This prevents a buildup of files from multiple uses of the tool and does not require a new file name each time the model is run. The Add to Display option will show the buffer in the map until the next time the model is run.

10 Right-click the Output Table variable and set it to add to the display also. Click the Full Extent button, if necessary, and save the model. The model is now fully in the Ready to Run state.

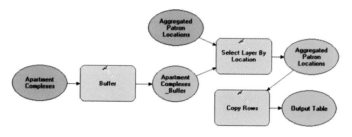

11 Use the Select by Rectangle tool on the Tools toolbar and select apartment complex number 56 in the map. You may need to move the model to the edge of the display to be able to see both the model and the map. Run the model either by clicking Model > Run Entire Model on the main menu or by clicking the Run button on the Model toolbar. The results are added to the map, and the model components now have a drop shadow to indicate that the model ran successfully. You may need to refresh the map window to display the selected patron locations.

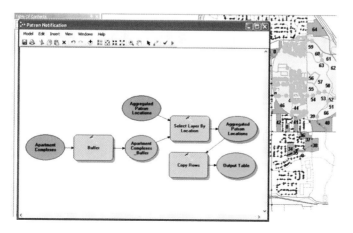

7–1

7–2

7–3

12 Click the List by Source button at the top of the table of contents, and then right-click and open the PatronLocation_Aggregated_Co file. This is the file that can be used to create a mail-merge letter and mailing labels. After examining the table, close it.

OBJECTID	Address	Number of Visits	City	Zip
1	1000 E ASH LANE #1104	1	Oleander	76039
2	1000 E ASH LN #1005	1	Oleander	76039
3	1000 E ASH LN #1308	1	Oleander	76039
4	1000 E ASH LN #207	2	Oleander	76039
5	1000 E ASH LN #710	1	Oleander	76039
6	1000 E ASH LN 413	1	Oleander	76039
7	1000 E ASH LN#1309	1	Oleander	76039
8	1000 E ASH LN. #1702	1	Oleander	76039
9	1000 E. AASH LN #2010	1	Oleander	76039
10	1000 E. ASH LANE #1512	1	Oleander	76039
11	1000 E. ASH LN #1705	1	Oleander	76039
12	1000 E. ASH LANE #1910	1	Oleander	76039
13	1000 E. ASH LANE #704	1	Oleander	76039
14	1000 E. ASH LN #1310	1	Oleander	76039
15	1000 E. ASH LN #1709	1	Oleander	76039
16	1000 E. ASH LN #809	1	Oleander	76039

(0 out of 918 Selected)

Add model parameters

The model is set to run as many times as you like, but the output file name and the buffer distance will stay the same. You could edit the model and change the name of the output file each time you want to run it, but that wouldn't be very convenient for the librarians who will be using this model. To make the model more flexible, you will set the buffer size and output file name as parameters of the model that can be set by the user. Once user-definable parameters are created, the tool can be run from either the Catalog or the Search window and a prompt will be generated to receive user input as with any other tool.

1 In the Model window, right-click the Buffer tool and go to Make Variable > From Parameter > Distance.

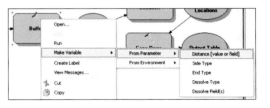

2 This adds a variable to the model for the buffer distance. Select the new Distance variable and drag it above the Buffer tool for clarity (you may have to clear the Buffer Tool icon first). Right-click the Distance variable and click Model Parameter. Notice that a large P is placed next to the variable to indicate that it is now a parameter.

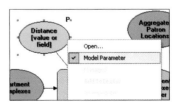

3 Repeat the process to set the Output Table as a model parameter. Save and close the model.

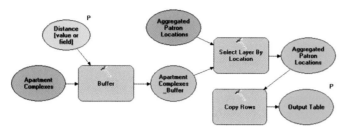

4 Go back to the Catalog window and expand the Library Models toolbox. Double-click Patron Notification to run it. You will see that the user is now prompted for the buffer distance and the name of the output table. Set the distance to **1320** feet and the output file name to **Complex_56**. Click OK to run the model.

The results are written to the user-defined file name, so now the model can be run multiple times without overwriting previous notification lists. Try selecting other apartment complexes or buffer distances and run the model to see the different results.

It is important to note that once you set variables in a model to be parameters of that model, you should not run the model from the model window. Only when the model is run from the Search or Catalog window will the prompt dialog box be displayed for users. Otherwise, the model will try to overwrite existing files and may generate errors.

5 Save your map document as **\GIST3\MyAnswers\Tutorial 7–1.mxd**. If you are not continuing to the next exercise, exit ArcMap.

7-1

7-2

7-3

Exercise 7–1

The tutorial showed how to create a simple model to replicate a geoprocessing task. The task involved having the model prompt the user for input that would affect how the model runs.

In this exercise, you will repeat the process and create a new model for the Public Works Department. When repairs are made to a street, the Public Works supervisor likes to mail out a notice to the people living on or near that street so that they can plan for the expected inconvenience. The supervisor would like you to automate the process with a model that will take a selected street centerline, buffer it by a user-defined distance, select the property owners within that distance, and generate a new table of addresses.

Next week, Park Crest Avenue will be repaved, so this will be the test case for the model.

- Start ArcMap and open Tutorial 7–1.mxd
- Go to the Exercise 7–1 bookmark.
- Create a new model in the Library Models toolbox called **StreetNotify**.
- The process will be as follows:
  - User selects a street centerline.
  - Buffer it by **100** feet (model parameter).
  - Use the buffer to select the property from Property Ownership.
  - Create a new table with the results (model parameter).
- Test the model on other streets with different buffer distances.
- Save your results as **\GIST3\MyAnswers\Exercise 7–1.mxd**.

WHAT TO TURN IN

If you are working in a classroom setting with an instructor, you may be required to demonstrate the models you created in tutorial 7–1 and exercise 7–1.

Printed screen capture of the model diagram from

  Tutorial 7–1

  Exercise 7–1

Tutorial 7–1 review

This tutorial examined the development of a simple model that would replicate a process done by the City of Oleander library. This model was developed based on several known requirements. Once translated into a GIS workflow, these requirements could be duplicated with ArcMap using **ModelBuilder**. The process as translated was to (1) identify the apartment complex, (2) conduct a buffer operation around the complex, (3) select the addresses of known library patrons within the buffered distance, and (4) output the addresses of the patrons to a table for later processing.

You began building the model by setting up the **environment** in which ModelBuilder operates. This was done to help identify which parameters to use when tools are connected in the model window and to set up the **default workspaces** to store the results.

With the help of ModelBuilder, you used four spatial operations in ArcMap to develop the Patron Notification tool. ModelBuilder does not add any new functionality to ArcGIS. It combines existing functionalities to simplify overall usage. Consequently, before you begin to build a model using ModelBuilder, you should first go through the steps manually to make sure that the desired process can be accomplished with ArcGIS tools.

When the tools were added to the model, they required **variables** to act as the input and output for the tools. This was accomplished by specifying the **output variable** from the buffer process as an **input variable** for the selection process. Then, the resulting patrons within a half mile were used as an input for the Copy Rows tool to go into the output table. In this manner, a complex string of tools can be created to perform several processes.

A P next to the variable in the model window designates that variable as a **parameter**. A parameter is a variable that requires user input to continue running the model. You created the parameters from the Make Variable > From Parameter menu. Once the model was run, the parameters prompted the user for the buffer distance and the name of the output table.

Visual cues indicate one of three different **states** as you built the model. The first state, indicated by the model components having no color, is the Not Ready to Run state. This is indicative of the lack of a required entry for one or more of the tools. Once all the required parameters are set, the model moves into the Ready to Run state, indicated by the addition of color to the model components. After the model successfully runs, the components are shaded with a gray drop shadow.

Once your model is complete, it can be run from the **model window in ModelBuilder** or from the Search or Catalog windows. Once you include model parameters to prompt the user for input, you should no longer run the model from the model window. Doing so will not generate an input screen, and the model will be run using only the default settings.

7–1

7–2

7–3

STUDY QUESTIONS

1. Why was it important to specify that ModelBuilder "overwrite the outputs of geoprocessing operations"? Can you think of a situation where this should not be the case?

2. Why was it important to specify the current and scratch workspaces during the initial environment setup?

3. What do the different colored "states" of the ModelBuilder components signify?

Other real-world examples

ModelBuilder can also help to automate the simulation of natural processes such as the impacts of storm water runoff. Variables such as land use, soil type, stream and flow characteristics, watershed areas, and rainfall amounts can all be simulated in ModelBuilder to arrive at the amount of storm water runoff. This amount is then coupled with average pollutant loads to estimate the pollutant impact of storm water runoff. Many agencies use such processes to estimate the annual pollutant loads on affected water bodies caused by storm water runoff.

Oil and gas drilling companies can use ModelBuilder to estimate the potential conflict such sites could have with surrounding urban development and environmentally sensitive areas. A proposed location for drilling can be examined with respect to zoning and land-use maps, residential areas, and environmentally sensitive areas such as wetlands, water bodies, and streams. Since the process needs to be done many times before a site can be open for drilling, ModelBuilder is an efficient way to conduct this analysis.

Tutorial 7–2

Expanding model capabilities

Simple models are fairly easy to develop using the model window in ModelBuilder. More complexity can be added by using the ArcGIS interface. Such capabilities include decision-making logic, adding different types of variables, and controlling the flow of the process.

Learning objectives
- *Define other variable types*
- *Substitute variables*
- *Set model preconditions*
- *Make decisions*

Introduction

The simple models that connect a string of existing tools are useful in automating a process. The model can manage intermediate data and display the output to the current map display. By using additional model capabilities, it is possible to add another level of complexity to a model.

The variables added in tutorial 7–1 were derived from components already in the model. They were basically substitutions for parameters that would be filled in if the tools were run outside the model environment. Models also allow for user-defined variables that can be substituted anywhere in the model process. These variables are shown as stand-alone components in the model and will prompt the user for input through a custom dialog box when the model is run.

When these types of variables are used independently, or sometimes when a model is asked to perform disparate tasks, there can be a problem with controlling the order of the processes. Models create a linear process and will follow the order in which the various components are connected. Having multiple processes in a model may cause it to run in an order not determined by the user. In that case, the model can be controlled with preconditions, which are restrictions placed on model components that delay their start until other conditions exist. This may be the existence of a file or the completion of another process.

7–1
7–2
7–3

As more complexity is added to a model, it becomes more like a visual programming language. A set of tools can be strung together, variables defined, and control over the order of processing maintained. A major component of any programming language, and one of the most powerful tools to use, is the ability to branch a process based on a tested value. This is commonly known as an if–then statement. A value is tested, and two options are presented based on the outcome of the test. For example, if a value were positive, it could trigger a certain process, whereas if the value were negative, another option might be used. The ability of the model to evaluate a situation and make a decision based on the outcome is a powerful programming tool.

As you become more familiar with the components available, you will be able to build models of increasing complexity that can handle a variety of situations.

Adding complexity to the model

Scenario   The librarians of Oleander are very pleased with the model you completed. They have a request for an enhancement, however. It seems that they are amassing a large number of output files and they would like to save the results for future reference by creating a new folder every time the model is run. Any of the five librarians may run the model in a given month, and they would like to have a personalized folder created each time the model is run. They intend to give the folders a name that reflects the librarian's name and the date the model is run. If the librarians run the notification for several complexes on a single day, all the files would be kept in that folder. When the bookmobile repeats a visit to a complex, the folder with the older date can be overwritten in favor of a new one.

Data   This tutorial uses the same data as tutorial 7–1 and builds upon the model you created.

Tools used   ArcMap:

ModelBuilder tool

ModelBuilder: In-line variable substitution

ModelBuilder: Python script

You will begin by adding a process to prompt the user for a new folder name and to create the folder. The new name will be stored as a freestanding variable, and then substituted into a process to create a folder.

Create a folder with a variable name

1 Start ArcMap and open Tutorial 7–2.mxd. Open the Catalog window and navigate to the Library Models toolbox. Right-click the Patron Notification model and select Edit. The model should appear in the Ready to Run state.

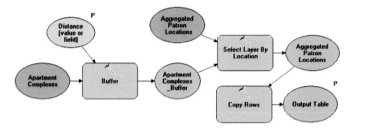

2 Add a stand-alone variable by clicking Insert on the main menu of the model window and selecting Create Variable.

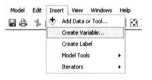

3 Select Any value. Before clicking OK, scroll through the list of possible variable types to get an idea of what selections are possible. Click OK.

4 Right-click the new variable and make it a model parameter. Right-click again, select Rename, and give the variable the name **New Folder**. Click OK.

5 Use the Search window to locate the Create Folder tool and drag it into the model to the right of New Folder.

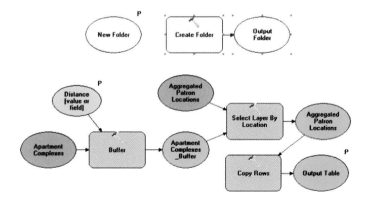

7-1
7-2
7-3

6 Double-click the Create Folder tool. Use the Browse button to set the folder location to MyAnswers. Click Add.

7 Under Folder Name, use the drop-down arrow and select New Folder. Because the variable is now a component of the model, it is available in the drop-down list. Click OK.

8 Right-click the variable Output Folder and rename it **New Library Folder**. Save the model. The new components appear in the Not Ready to Run state. This is because the value for the New Folder variable has not been entered and can only be populated when the model is run from the Search or Catalog windows. Remember that once you start working with model parameters, the model should no longer be run from the model window.

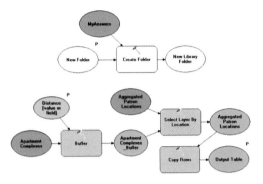

Implement in-line variable substitution

The new process added will create the desired folder, but now the folder needs to be combined with a name for the new table and added as the name and location for the output table. A second stand-alone variable will be needed for the table name, and then it will be combined and substituted for the output file name in the Copy Rows tool.

1 Go to Insert > Create Variable. Set its type to Any value and click OK. Rename it **Mail List Table**. Since the user will need to be prompted for this value, right-click the variable and set it as a model parameter. The variables have been created for the new folder and the new output table. To make these the output location and file name used by the Copy Rows tool, you'll use a technique called in-line variable substitution. By enclosing the variable name in percentage signs, you will tell the model to substitute the value of the variable in the input box. Several variables can be included on the same line provided that when the values are substituted, they constitute all that the command will need to execute.

2 Double-click the Copy Rows tool to open its dialog box. Change Output Table to read

%New Library Folder%\%Mail List Table%.dbf

This would make a path that includes the new folder name, a backslash, the new table name, and the correct file extension. Click OK.

3 Pause the cursor over the Output Table variable and note the path and file name. Since this input is now being handled by other variables, right-click the oval and clear the check box for Model Parameter. At this point, you could try running the model, but there is no control over which of the processes will run first.

What if the buffer/selection/copy rows process runs before the create folder process? The folder would not exist yet and the copy rows process would fail. The solution is to set a precondition on the Copy Rows tool that will not let it execute until the variable New Library Folder exists.

4 Right-click the Copy Rows tool, and then click Properties. Click the Preconditions tab and select the New Library Folder check box. Click OK. Notice that the model now has a dashed line connecting the New Library Folder variable and the Copy Rows tool. This signifies that the Copy Rows tool will not execute until the New Library Folder variable exists. Note that you can click connector lines to add vertices and have them go around model elements to make your model look neater.

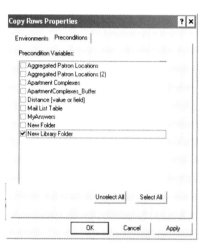

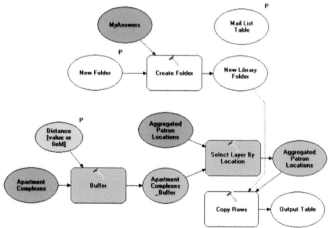

7-1

7-2

7-3

5 Save the model and close it. Make sure an apartment complex is selected in the map, and then double-click the Patrons Notification model to run it. Enter a new folder name of **Marion** and an output table (Mail List Table) name of **Complex_56**. Click OK to run the model.

When the model completes, a pop-up box will appear to the lower right of your screen. When opened, it will display the input and environment values used to run the model and any messages that the model generated, including errors. If the pop-up box dissolves before you open it, you can open it from the main toolbar by going to Geoprocessing > Results.

6 Click the pop-up box to see the results, and close the box after you've reviewed them.

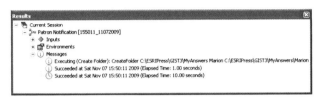

7 Because the final output table created by the Copy Rows tool was set using in-line variable substitution and not as a model parameter, the model cannot automatically add it to the display. Click Add Data, browse to the MyAnswers\Marion folder, and add the table Complex_56 to the table of contents. The table will hold the results, but the important thing to note is that a new user-defined folder and file name were created from the variables you defined. If you were to have several people running this tool multiple times, you would eventually discover a problem with the process. The model is designed to run with a single feature selected. If the user forgets to select a feature, the buffer command will treat this as though all the records are selected. This will result in dozens of buffers being created and practically all the library patron features being selected, not to mention the extended time it will take for the processes to finish.

To tackle this second problem, you'll add a tool to count the number of selected features. Then you'll have the model execute a script that will look at the number of features and make the decision whether or not to continue.

Add decision-making capabilities

1 Right-click the Patron Notification model, and then click Edit to open the model window. Search for the Get Count tool and drag it into the model near the Apartment Complexes input variable.

2 Use the Connect tool to set ApartmentComplexes as the Input Rows value for the Get Count tool. Click OK. The output will be a variable called Row Count, which will be used to show how many features are selected. A script is provided in a custom toolbox that will take the row count and see how many features are selected. If any number other than 1 is

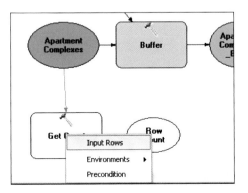

selected, the script returns the Boolean value False. If a single feature is selected, the script returns the Boolean value True. While it would be beyond the scope of this tutorial to have you write Python scripts, an attempt will be made to describe how the code works.

3 **In the Catalog window, navigate to C:\ESRIPress\GIST3\Scripts and find the Test Count Value script. Right-click the script and then click Edit. The Python editing interface will open. Follow along in the script window and review the code.**

```
# Name: CountTest.py
# Author: David W. Allen, GISP
# Check the input value to see if it is greater than 0.
#
# --------------------------------------------------------------

# Import arcgisscripting and sys modules
import arcgisscripting, os, sys
gp = arcgisscripting.create(10.0)
```

The first segment brings the ArcGIS geoprocessing tools into the Python framework. Any command prefaced with gp. will be pulled from these tools.

The second part defines a variable that will accept the input values for the script. This will be the output variable from the Get Count tool in the model.

```
# Set up input parameter for use in a model
InputVal = gp.GetParameterAsText(0)
```

The next part sends a message to the results dialog box letting the user know how many features are currently selected. This also includes a line to set the output parameter to False.

```
# Report back how many features are found
gp.addmessage ("The number of selected features = " + InputVal)
gp.SetParameterAsText(1,False)
```

The last part looks to see if the number of selected features equals 1. If it does, the output variable is set to True and a message is sent to the results dialog box. If any other number of features is selected, the output variable remains False and an appropriate message is displayed.

```
# Set the output to True if there is only one feature selected
if InputVal == "1" :
    gp.SetParameterAsText(1,True)
    gp.addmessage ("Correct number of features to continue running.")
else:
    gp.addmessage ("Either no features or too many features were selected!")
```

4 **Close the Python editor windows. To test the script, double-click to open it and enter a value of 1. Click OK to run it.**

5 **Open the Results window to see the messages. Note the message reporting that the correct number of features is selected to continue. Try again with any other number and note the messages. Close the Results window when you are done.**

7–1

7–2

7–3

6 Drag the Test Count Value script into the model next to the Row Count variable. Use the Connect tool to connect the Row Count variable to the Test Count Value script and set the input as InputVal. The output of the Test Count Value script is Boolean—either True or False. A precondition set on the Buffer tool would prevent it from running unless the value of the OutputVal is True.

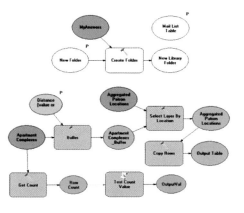

7 Right-click the Buffer tool, and then click Properties. Click the Preconditions tab and select the OutputVal check box. Click OK.

YOUR TURN

You would not want the model to create the requested new folder unless the correct number of features were selected for the model to run. To control this, set OutputVal as a precondition for the Create Folder tool. This will make the OutputVal variable a precondition for both the Create Folder tool and the Buffer tool. Since it is no longer necessary to have OutputVal as a precondition for the Copy Rows tool, select the line representing the precondition in the model diagram and delete it. Note that you can also set a precondition with the Connect tool. Then the two model components are connected, and one of the choices in the pop-up dialog box is to make the connection a precondition.

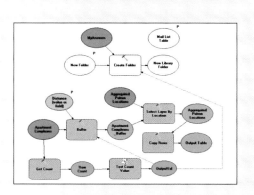

8 Clear all selected features and run the model. Set the new folder name to **David** and the output table to **Complex_00**. When the model completes, click the Results pop-up box. Note the message from the script stating that no features were selected.

9 Select an apartment complex in the map area and run the tool with the same folder name and the complex number of the polygon you selected. When the process is complete, investigate the folder by using Add Data to add it to the table of contents and see if the new output table is in it. Now, the model has more flexibility with the addition of stand-alone variables and in-line variable substitution as the inputs to a process to create a new folder for the results. It also has some error checking in the form of preconditions to make sure the processes run in the correct order. Finally, it has decision-making capability to determine whether the correct number of features is selected in order for the model to finish successfully.

7-1

7-2

7-3

10 Save your map document as **\GIST3\MyAnswers\Tutorial 7–2.mxd**. If you are not continuing to the next exercise, exit ArcMap.

Exercise 7–2

The tutorial showed how to create stand-alone variables and substitute them in the parameters for other processes. It demonstrated the use of preconditions and included a script to make decisions about running the model.

In this exercise, you will add some of the same features to the model you created for the Public Works Department in tutorial 7–1.

- Start ArcMap and open Tutorial 7–2.mxd.

- Start editing the StreetNotify model.

- Add a stand-alone variable for a new table name, and build the process to create a new folder.

- Use in-line variable substitution to define the final output table name.

- Set a precondition to prevent the model from writing the final output table until the new folder is created.

- Add the Get Count tool to determine the number of selected features.

- Review the code for the Street Count Test script. Note that it will allow as many as twenty features to be selected and still run.

- Set up a precondition on the Buffer tool to not run if too few or too many features are selected. Use the Street Count Test script in the provided toolbox.

- Change the preconditions for other tools to keep them from executing if the correct number of features is not selected.

- Test the model.

- Save your results as **\GIST3\MyAnswers\Exercise 7–2.mxd**.

WHAT TO TURN IN

If you are working in a classroom setting with an instructor, you may be required to demonstrate the models you created in tutorial 7–2 and exercise 7–2.

Printed screen capture of the models created in

       Tutorial 7–2

       Exercise 7–2

Tutorial 7–2 review

This tutorial added additional complexity to the model from the previous **ModelBuilder** tutorial (tutorial 7–1). This functionality led to further enhancements once the initial effort is complete. In the case of this tutorial, you used the functionality of ModelBuilder to enhance the customer's workflow by creating a process that would build tighter organization into a series of processes. This was achieved by building several new **functions** into your model.

First, you added a **stand-alone variable** in the model that would create a new folder to store the results of the Patron Notification tool. A **parameter** for this variable was defined so that the name of the resulting folder could be customized to the individual conducting the model.

A second stand-alone variable was created to contain the new output table. The **in-line variable substitution** technique was used to specify that the table be created in the new folder the user creates and that the table be named according to the apartment complex selected for the analysis.

You added an error-checking mechanism to ensure that the model does not write the final output table until the new folder is created. This is called **setting preconditions** and is used to make sure that all the processes contained in the model operate in the correct order.

Stepping through the logic of the model, it became apparent that you needed a way to ensure that it was operating as intended with respect to the number of records being selected, so you used a **Python** script to build a tool to count the number of selected features. This tool provided **decision-making logic** in the model, so that depending on the result of the count tool, the model would either continue running or stop.

STUDY QUESTIONS

1. What is the purpose of adding a parameter to a variable?
2. What is in-line variable substitution, and why did you use it in your model?
3. What benefit is derived from having decision-making capabilities in your model?

7-1
7-2
7-3

Other real-world examples

A transportation department might use ModelBuilder to build a model of traffic flow. A model with user-defined variables would have the flexibility to simulate traffic at different times of day to accommodate varying traffic loads. Staff would also be able to adapt the model to a weekday versus weekend study by entering different traffic parameters.

Scientists studying critical habitat for endangered species might build a model to track the change in habitat. The size or condition of the habitat may be controlled by user-defined variables. This would help fine-tune the suitability rating of various habitat types with input based on field observations.

A custom toolbar might be built to contain several models pertaining to a certain study. For example, a county engineer studying groundwater contamination might need to develop several models to track different aspects of the groundwater. All the models could live on a custom toolbar and be readily available to run when needed.

Tutorial 7–3

Creating model documentation

As you create models, you will realize that they can be made to operate like the standard tools available in ArcGIS. These custom tools will not only make your models look good, but they can provide a rich Help environment that will help users to understand how to use them more effectively. This is done by building documentation inside both the model editor environment and the user interface dialog box.

Learning objectives

- *Configure stand-alone labels*
- *Arrange model components*
- *Configure component-linked labels*
- *Build context-sensitive Help*

Introduction

Extensive customization can be done with ModelBuilder. One of the most important things you can do is thoroughly document your models for the benefit of users as well as help identify how the customization has been done in the event that future modifications need to be made.

The model editor dialog box lets you do two types of documentation, which appear in the model diagram. One type of labels is linked to the components of the model. These labels can be used to describe variables and tools, and this type of label moves when the model element is moved. The other type of label is not linked to any model component and remains in a fixed location unless it is specifically moved.

Models can also have custom Help files created that will provide descriptions and examples of how the model is used. These can be in the form of general descriptions, context-sensitive Help, images, examples, bullet lists, and much more. These are built in a rich Documentation Editor environment. Once this type of documentation is created, it will be hard to distinguish the user interface and available Help for a custom tool from the user interface and Help for a standard ArcGIS tool.

Building the documentation

Scenario The model for the library staff turned out very well, and they are all very happy with it. Because it will only work if a single feature is selected, and there are user inputs that require explanation, a Help environment needs to be built. You'll start with labels to help make the model diagram easier to understand, and then build a Help environment for the tool.

Data The work in this tutorial will build on the model completed in tutorial 7–2.

Tools used ArcMap:

> ModelBuilder tool
> ModelBuilder: Documentation Editor

Label the model

1 Start ArcMap and open Tutorial 7–3.mxd. Open the Catalog window if necessary and navigate to the Library Models toolbox. Right-click the Patron Notification model and click Edit. This is the completed model, but anyone else looking at this model diagram would not understand what the components do.

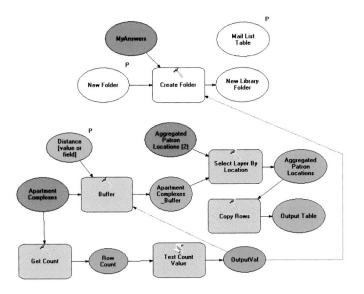

2 All the work with labels will be done using the Select tool on the Model toolbar. Click the Select tool. Then go to Insert > Create Label on the main menu. You'll see a new label called Label added to the display.

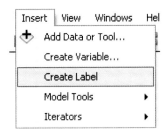

3 Right-click the new label, and then click Display properties.

4 The resulting dialog box can be used to set a variety of properties for the label. Clicking in any of the boxes will allow the parameter to be changed. Click in the Name box and change it to **Patron Notification**. Many of the parameters have an additional dialog box that can be used to change their setting. This is accessed by clicking the line you want to change, and then clicking the button that appears on that line, which is called the ellipsis button.

Display Properties	☒
Selected Label Properties	⌄
Name	Patron Notification
Tooltip	
URL	
Font	MS Shell Dlg
Text Justification	Center
Resizability	Tight Fit
Background Color	
Border Color	■
Border Width	
Transparent	True
Width	39
Height	16
X Center	38
Y Center	359
Region	Any
X Absolute	39
Y Absolute	359

5 Click the Font box, and then click the ellipsis button at the right of the box. Change the font size to **20**.

YOUR TURN

In the Display Properties dialog box, click each of the rows listed here and use their ellipsis buttons to change their setting:

- Background color to Light Fuchsia
- Border Width to the largest choice

Close the properties dialog box and move the label to the upper left of the diagram.

7-1

7-2

7-3

Work with graphics in a model

Although ModelBuilder doesn't have any graphic drawing tools, it would be nice to have a box around some of the components to highlight what they do. This can be done with a label.

1 Go to Insert > Create Label to add another label. Set its parameters as follows:

- Name: Test for number of items selected
- Border Color: Blue
- Border Width: the thinnest line
- Font size: **14**

2 Drag the resulting label to the area of the diagram that contains the script for checking the number of features. You may need to zoom out a little to get the label placed.

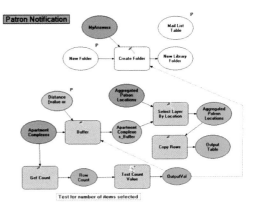

Create similar labels for the processes that create the new folder and perform the buffer/selections. Use a different color for each box, but keep the text size the same. You may want to move the model components a bit to make room for the labels. Clean up the connection lines as necessary.

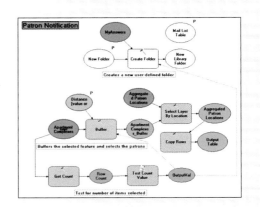

Export the results

1 To save the model diagram to a graphic file, go to Model > Export > To Graphic on the main menu.

2 In the resulting dialog box, you are able to save the image to any number of formats. Set the output format to JPEG. Use the Browse button to find MyAnswers and name the file **Patron Notification Model Diagram**. Click OK to close the browser, and then OK to perform the export. Save and close the model.

3 **The resulting diagram can be used to show others what each component of the model does.** As you can see, there are many ways to dress up the model diagram for future reference. It doesn't help end users very much, however, since they are not likely to open the model window, or even understand what they would see there.

They need to have messages and explanations of how things work available at the user interface. There are many levels of documentation that can be built for this purpose. Step one will be to investigate an existing tool to see how Help functions.

Examine an existing tool

1 Use the Search window to find and open the Intersect Tool dialog box. If necessary, click Show Help at the bottom of the dialog box to reveal the Help notes. Here you will see a general description of how the tool functions.

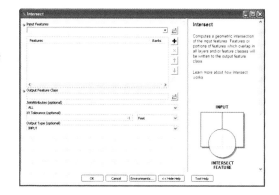

7-1

7-2

7-3

2 Click in the Input Features box. Notice that the description in the Help window changes to reflect the area of the dialog box you selected. Try a few other areas and note how the description changes. This is called context-sensitive Help because it changes according to the dialog context.

3 Finally, click Tool Help. A rich dialog box containing tool descriptions, illustrations, script examples, and command-line help appears, along with the search window of ArcGIS Desktop Help. Scroll through the documentation window and examine the types of components shown. They include tables, images, bullet lists, and more. Once you are done examining ArcGIS Desktop Help, close the dialog box and cancel the Intersect tool.

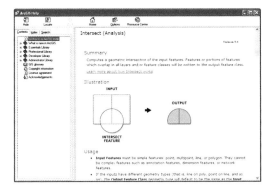

4 Examine Help for other tools if you wish to see how their Help files are organized. When you are done, close ArcMap and save your map document as **\GIST3\MyAnswers \Tutorial 7–3.mxd**.

All of what you see can be built for a custom tool through the documentation dialog box. Once you have built this level of documentation, it is stored in the ArcGIS Desktop Help files and is fully searchable by any user. Someone totally unfamiliar with the tool could search for the word Patrons in ArcGIS Desktop Help and your custom tool would appear in the search results.

Build user interface documentation

The first part of the documentation to build will be for the user interface. This is what appears in the tool dialog box without clicking the Tool Help button.

1 Open ArcCatalog and navigate to the MyToolboxes\Library Models toolbox. Expand the toolbox to show the models it contains.

2 Click the Patron Notification model, and then click the Description tab. This dialog box can be used to edit the description or view the metadata. Your tool should have several areas in the Help file to explain how it is used. The first will be a general description that will appear when the tool is first opened. It should have a general summary showing what the purpose of the tool is and remind the user that a single feature must be selected. It should also have a bullet list showing what data is necessary for the tool to function. Finally, it should contain an illustration of the model diagram. In addition, each of the input parameters should have its own Help file to describe what data constitutes valid input.

3 Start by clicking the Edit button in the upper left of the pane. In the Summary area, type the following:

> **The tool will buffer a selected apartment complex by a specified distance, select the library patrons within that distance, and write a new table containing the results.**

4 In the Image area, click Update under the sample image. Navigate to the Data folder and select Patron Notification Model Diagram.jpg.

5 Click Tags. The keywords are what the Help search will look for to determine when the tool should be included in a set of search results. Add the keywords **Library**, **Patron**, **Notify**, and **Notification**, all separated by a comma. The Description area allows you to enter customized text describing the tool. This can include images, bullet lists, or Web links and can be formatted with different fonts, justification styles, colors, and more.

7–1

7–2

7–3

6 In the Description area, click the Bullet List button and type **User must have a single apartment complex selected**. Press enter and add another bullet to read **If a single apartment complex is not selected, the tool will stop**. Add one more bullet to read **The Apartment Complex data and recent Library Patron data are necessary for the selection to complete successfully.** Press Enter twice after the last bullet to end the list. The Parameters area has the same rich set of formatting tools as the Description area and lets you enter Help tips for each of the requested user inputs. The parameters shown in the accompanying diagram are the ones you set as model parameters and saw in the tool dialog box when you ran the tool.

7 In the Parameters area in the Explanation pane for the Distance_value_ or_field_ parameter, type the text **The user can define the search distance for selecting Patrons**. Press Enter to start a new line, and type **Enter the search distance for notification**. Highlight the last line of text and click the Italics button.

YOUR TURN

Enter similar explanatory text for the next two parameters as shown:

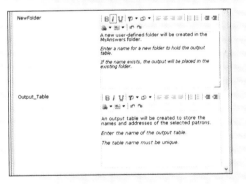

New_Folder

A new user-defined folder will be created in the MyAnswers folder.

Enter a name for a new folder to hold the output table.

If the name exists, the output will be placed in the existing folder.

Mail_List_Table

An output table will be created to store the names and addresses of the selected patrons.

Enter the name of the output table.

The table name must be unique.

As you can see, the combination and complexity of elements is endless. Add a few more if you like, or try some of the other format settings.

8 Copy and paste the same information into the Explanation pane under Scripting syntax.

9 Save the updates. Close ArcCatalog.

Examine the results

1 To see the results, start ArcMap with tutorial 7–3, navigate to the Patron Notification model, and run it. Notice the tool description in the Help dialog box, and then click the Distance input box and watch how the Help documentation changes. Try clicking in the other input boxes to see the changes.

2 Click Tool Help and examine the rich dialog box that has been built. Scroll to the bottom and notice the model elements table. When you are finished looking, close Help and cancel the tool.

3 Click the Search tab and type **Patron**, which was one of the keywords entered earlier. Click Search. Your custom tool has become part of the ArcGIS fabric! The Help documentation you created is remarkably similar to the Help provided for standard tools. Here's a secret—the ESRI developers build it exactly the same way.

4 Save your map document as **\GIST3\MyAnswers\Tutorial 7–3.mxd**. If you are not continuing to the next exercise, exit ArcMap.

7–1

7–2

7–3

Exercise 7–3

This tutorial showed how to develop a rich documentation and Help environment for the Patron Notification model. There are components that are displayed in the model diagram, the context-sensitive Help, and the Tool Help dialog box.

In this exercise, you will build a similar documentation framework for the Public Works street notification model from exercise 7–2.

- Start ArcMap and open Tutorial 7–3.mxd.
- Create labels, graphic boxes, and clean precondition lines in the model diagram.
- Add documentation to the tools in the diagram.
- Save a copy of the finished diagram for use in Help.
- Fill in general information for the Help documentation.
- Fill in the Help components in the documentation.
- Save your results as **\GIST3\MyAnswers\Exercise 7–3.mxd**.

WHAT TO TURN IN

If you are working in a classroom setting with an instructor, you may be required to demonstrate the models you created in tutorial 7–3.

Printed screen capture of the model diagram from

> Tutorial 7–3
> Exercise 7–3

Tutorial 7–3 review

Extensive customization can be done with **ModelBuilder**. Properly documented models built with ModelBuilder are hard to distinguish from the core ArcGIS tools. To achieve this authenticity, you must thoroughly document your models for the benefit of users as well as to explain how the customization has been done in the event that future modifications need to be made.

In this tutorial, you documented the model that you created in the earlier two ModelBuilder tutorials (tutorials 7–1 and 7–2). First, you added explanatory graphics to the **ModelBuilder view**. The diagram serves as a reminder of how the model works, but more importantly, it also would help someone better interpret the model if they had not seen it before. **Graphic elements** such as labels and boxes that group objects together help **clarify the intention and flow of the model**. You exported the model to a graphics file so that you could print and keep a hard-copy record of your model.

After reviewing an example of documentation for an existing ArcGIS tool (what better way to measure how thoroughly you are documenting your model), you proceeded to the next level of documentation—building **user interface documentation**. The user interface documentation is what appears in the tool dialog box without clicking the Tool Help button.

The **ArcCatalog Description Editor** in ArcCatalog was used via the Edit button to document the **Description**, **Command line syntax**, and **Scripting syntax** categories. Information typed here not only documents the model, but also appears in the **Tool Help** area when users look for clues on how to operate the model.

STUDY QUESTIONS

1. How is documentation added to the model window?
2. What options are available for formatting your version of Help in ArcCatalog?
3. What three components of Help should you make sure your tool has documented?

7–1
7–2
7–3

Other real-world examples

A transportation department simulating traffic patterns with a model may want to share the model with other agencies. Building a clear set of instructions and context-sensitive Help would allow the other agencies to quickly understand the model and its parameters. Then they could adapt the model to their own circumstances and observations.

Scientists routinely share their analysis techniques to promote advances in their fields. The study of habitat for endangered species is a global concern, so documenting any models that are shared may include adding Help screens in different languages. This would facilitate the rapid deployment of a model to keep up with rapid habitat change.

A state forestry agency might rely on state or national firefighters to combat a large forest fire. Good model documentation would help firefighters and volunteers become familiar with the analysis techniques of local agencies as well as track the fire's path. Using a model would lessen the learning curve and allow the staff to become effective much more quickly.

References for further study

There are a number of additional resources available to help with ModelBuilder. **ArcGIS Desktop Help** lets you use keywords to search for a title or topic. You can access ArcGIS Desktop Help in ArcMap or ArcCatalog by clicking Help on the main menu. Then click the Search tab, type a keyword, and click Ask. Use the keywords provided here to search for additional learning resources on many of the concepts taught in this chapter.

ESRI has developed a number of online courses on a wide variety of pertinent GIS topics. **ESRI Self-Study (Virtual Campus) courses** are an excellent resource for students that will supplement information covered in this course. ESRI Virtual Campus courses can be accessed at `http://training.esri.com/gateway/index.cfm`.

ArcGIS Desktop Help search keywords

ModelBuilder, model parameters, variables, models, customizing the interface, custom toolbar, context menu, toolbox, scripts, ToolTips, MapTips, custom dialog boxes, customizing, new menu toolbar, adding tools, programming tools, Python, programming languages, macros

ESRI Self-Study (Virtual Campus) courses

The following ESRI Virtual Campus classes may be helpful:

1. Geoprocessing Using ModelBuilder
2. Customizing ArcGIS Desktop
3. Customizing ArcMap: Easy Ways to Extend the Interface
4. Getting Started with Scripting in ArcGIS 9

Part 4
Using advanced techniques for labeling and symbolizing

Developing labels and annotation

Tutorial 8–1

Labeling with Maplex

Labels in a map are generated dynamically using an attribute of the features being labeled. For years, the process of creating labels in digital mapping was greatly hampered by the capabilities of the software. Labels, which are placed on the map according to user-defined rules and often done manually, had to be regenerated each time the map was zoomed or panned. The limited number of rules paired with limited computer speed could make this a trying task and produce less than ideal results. The addition of Maplex to ArcMap allows for a rich set of label placement rules, and faster computers make dynamic label placement a more acceptable process than before.

Learning objectives

- *Activate Maplex*
- *Choose label styles*
- *Set up label classes*
- *Prioritize labels*

Introduction

The Maplex Label Engine is fast becoming the tool of choice for dynamic label placement. The Maplex engine has the capability of considering your settings for label placement and matching them against the placement of other labels and existing map features. This keeps labels in their correct places while still making them clear and legible. The three areas of user-defined rules are Label Position, Fitting Strategy, and Conflict Resolution. It is important to note that all labels will be placed automatically by Maplex, and the rules you set will be used to make the labels appear in the best placement. However, there will still be labels that either cannot be placed, or their placement is not optimal. The trade-off is that one set of Maplex rules for labeling can produce acceptable results for a variety of map scales.

The label position determines how the label will be placed in reference to the feature it is representing. Each type of feature, whether point, line, or polygon, has its own set of choices. The user can define an optimum orientation to the feature, and Maplex will try to place the label in that position. If the label will not fit at the optimum location, Maplex will look at the user's hierarchy of placement and try to get the best placement possible.

For very large labels or a tight labeling position, the user may define a fitting strategy, which may alter the properties of the label to fit in the optimum placement. This strategy may include making the label smaller, narrower, or possibly abbreviating the text. The fitting strategy works with the placement rules to determine the location of the label.

The conflict resolution rules take the well-placed label and look for interference with other labels and features in the map. A weight value may be given to a label that will cause it to always appear on top of certain features, or always move it to make a feature visible. There are also settings for how often a label should be repeated alongside a feature, or if it should appear as a background label, which would allow features to draw over it.

All the Maplex rules are user definable, as is the order in which they are applied. The user may decide which things to try first to get a label to fit in the optimum position, and then control the order of the alternative placement rules. The goal is to have Maplex do every-thing possible to place every label automatically and reduce both the number of unlabeled features and the amount of manual work that the user must do on the map.

These three categories are set within a single layer, but there are also settings to control label conflict among layers. The Label Priority Ranking controls which labels should be placed first and not be moved to accommodate other labels. The other control is the Label Weight Ranking. These settings determine which features can cause labels to be moved in order not to obscure the features. The combination of these settings gives you another level of control over label placement.

Adding dynamic labeling

Scenario The maps that have been produced for the City of Oleander have never been great in the past. The former GIS manager used the ESRI Standard Label Engine with limited success. There have always been streets that didn't get labeled and ambiguities with many of the fea-tures because of poor label placement. As a result, a lot of hand labeling was done for the maps. The word has come down from the city manager that all the maps for Oleander need to be reworked for publication in a new *Developer's Map Atlas*. The atlas will need to include detailed maps to help lure new development to Oleander, so you will need to produce the best cartographically correct maps of your life.

The first part will be to make a map that developers will use to identify property. It will need to have the subdivision name, the block numbers, and the lot numbers. On top of that, you will overlay the sewer line information showing each line's material and size, as well as the depth to the flowline of each manhole. The maps will be 11-by-17 inches at a scale of 1 inch = 200 feet.

Data All the editing and data creation has been completed. You will concentrate on labeling the property and sewer lines for a presentation map.

8–1

8–2

Tools used ArcMap:

Maplex extension

Reference scale

Labeling toolbar

Maplex Label Engine

Label Manager

Abbreviation Dictionary

Set up the Maplex environment

Maplex is loaded as an extension in ArcInfo, so it must be turned on.

1 **Start ArcMap and open Tutorial 8–1.mxd. The map opens with the basic data showing.**

2 **On the main menu, go to Customize > Extensions. Select the Maplex check box if necessary. If Maplex is not loaded, check with your system administrator to get it loaded. Close the Extensions box.** The desired output maps for this project will be at a scale of 1 inch = 200 feet, or 1:2,400. In order to get the best label placement for this scale, you need to set the reference scale for the map document. Without setting a reference scale, Maplex will rescale the labels each time the scale changes from zooming in or out.

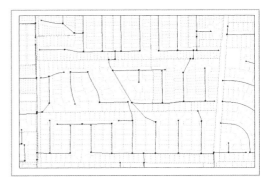

3 **Right-click the data frame name, Oleander Sewer Maps; point to Reference Scale; and click Set Reference Scale. This will accept the current scale as the reference scale.**

4 Next, go to Customize > Toolbars on the main menu and add the Labeling toolbar to your map document. The one shown here is undocked for clarity, but you may

dock it if you wish. The first button on the Labeling toolbar opens the Label Manager dialog box. The dialog box lets you manage all the labels for all the layers in the table of contents from a single screen, eliminating the process of opening the layer properties of each layer separately. The next two buttons set the ranking and priority of labels, deter-mining which labels will be drawn first and which labels may bump other labels out of the way. Next are two buttons to control the regeneration of labels, which can be time-consum-ing. Lock Labels will freeze the current set of labels while you work on other things, letting you see labels without having to wait for the regeneration process. Pause Labeling will turn all the labeling off, which means that you can work on other things but not see any labels at all. View Unplaced Labels shows the labels that Maplex was not able to place so that you may review your rule selections to try and include them. Finally, you have the option of setting the Maplex drawing mode to Fast or Best. Set it to Best while you are working on critical label placement, and then switch to Fast while you work on other parts of your map. Just remember to set it back to Best when the final maps are produced.

5 On the Labeling toolbar, click the Labeling drop-down arrow and select Use Maplex Label Engine. This document will now use Maplex Label Engine for all the labeling tasks. If you share this map document with someone who does not have Maplex, the labels will automatically be converted back to the Standard Label Engine and lose many of the labels because of limited placement rules.

Before moving on, take a look at some of the options in the basic label settings.

6 Click the Labeling drop-down arrow again and select Options. Click the General tab. Here you can set the color of unplaced labels, mark whether or not labels will rotate with the map, and designate the break angle for label orientation. Leave these at the default settings.

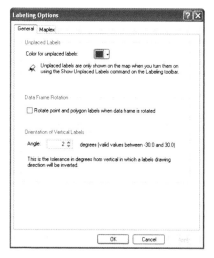

8–1

8–2

7 Click the Maplex tab. Here you can determine if line segments representing the same feature can be labeled as one, if the labels can flow outside the map border, or how to handle the labeling of multipart polygons. In this map, the sewer lines need to be labeled as one long pipe segment. Clear the "Enable connection of line segments into continuous features" check box. Click OK to close Labeling Options.

Label the plats layer

1 On the Labeling toolbar, set the mode to Best, and then click the Label Manager button.

2 Select the Base Map Group: Plat_Index check box and right-click Default. Select Rename Class and change the name to **Subdivision Numbers**. Click OK to close the rename box.

3 With the Subdivision Numbers label class selected, its labeling parameters are displayed. This screen is used to manage all the user-defined settings for Maplex. Set Text String to SubID.

4 In the Text Symbol area, click Symbol. Set the color to Seville Orange, the font to Arial, size to 20, and click both the Bold and Italic buttons.

5 Click OK to close the Symbol dialog box. Click Apply in the Label Manager, and you can see where the labels are being placed by default. With the desired symbol set, you can start to work on placement. The labels are rather haphazard in their default positions and overlap in some cases. A more organized placement would be to put the label in the corner of the plat boundary, shown with orange lines. These are also background labels, meaning that it is OK to place other labels over them if need be.

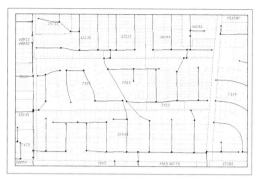

The Placement Properties area in the Label Manager dialog box contains a few of the most commonly changed settings for convenience, but to access all the settings, you need to click Properties.

6 In the Placement Properties area, click Properties. Click the Label Position tab. Leave the placement mode set to Regular Placement and select the "Place label at fixed position within polygon" check box. This will let you place the plat index number at the corner of the polygons and determine the order in which the placements will be tested for fit.

7 Next click Internal Zones. The numbers entered here determine which location will be used first for label placement. If the number 1 placement does not work for a particular label, Maplex will try to fit it at the next highest choice, then third, then fourth, and so on until it is placed. The top priority will be given to the upper-right corner of the polygon, and then the others as shown in the accompanying diagram. When you have the numbers filled in, click OK to close the dialog box.

8–1

8–2

8 Click the Fitting Strategy tab and select the Reduce font size check box. Click Limits and you will see that this rule can reduce both the size and compression of the font. Click OK to close the Label Reduction dialog box. Clear the check boxes for any other rules on the Fitting Strategy tab.

9 Click the Conflict Resolution tab. Set these labels as background labels by selecting the check box for this rule.

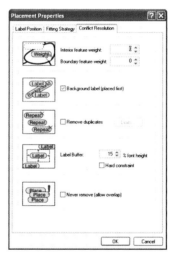

10 Now go back to the Fitting Strategy tab and click Strategy Order. This will let you set which of your defined rules will be applied first. The first rule should be to try reducing the label size to make it fit, and then try compressing the letters together. Select the rules from the list and use the arrows at the side of the dialog box to move them to the top of the list. Click OK to close the Strategy Order dialog box.

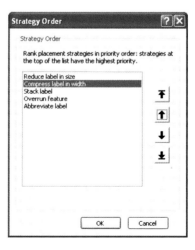

11 Click OK to close Placement Properties, and then OK again to close the Label Manager. As you can see, Maplex did a pretty good job of placing all the labels in the corners of the polygons where they will be easily visible. If these did not turn out well, you could go back to Placement Properties and tweak them to get a better result.

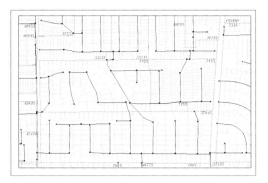

Test each trial by clicking View Unplaced Labels to see how many labels could not be placed using the current rules.

Label the block numbers

Property is identified by its subdivision index number, block number, and lot number. You've already placed the subdivision index numbers. The block numbers will come from a set of polygons that represent each block. They have an attribute called Block Designations that will be used for the labels.

1 Turn on the Blocks layer, right-click, and then click Properties. Next click the Symbology tab. You will see that the blocks are symbolized with a solid blue color.

2 Click the Symbol preview to open the Symbol Selector. Set the color to No Color and the outline width to 0. This means that the geometry for the blocks will not be shown at all. Click OK, and then OK again to accept the changes.

3 Open the Label Manager dialog box and select the Base Map Group: Blocks check box. Then click Default to set the label settings. Verify that Text String is set to Block Designations.

4 There is a special symbol for block designations that will enclose the block number in a circle. In the Text Symbol area, click Symbol. Type **Block number** in the search area and locate the ESRI block number symbol. Click it, and then click OK.

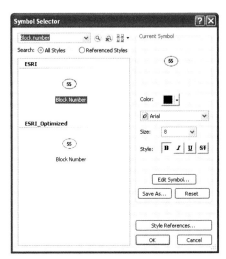

5 Click Apply and drag the Label Manager out of the way to look at the label placement. In the center of the map, there are a few blocks that have the block number repeated. These need to be fixed.

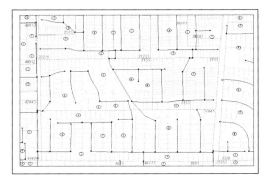

6 In the Placement Properties area of the Label Manager, click Properties, and then click the Conflict Resolution tab. Select the Remove duplicates check box and click Limits. Set the search radius to **200** map units. Click OK to accept the distance, and then OK again to close Placement Properties.

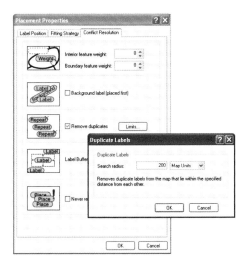

7 Click Apply and move the Label Manager out of the way to see the results. The problem is resolved. Close the Label Manager.

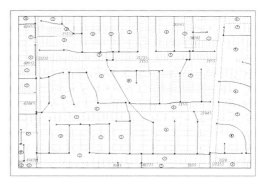

Label the lot numbers

The layer Property Ownership will be used to label the property with the lot number. Like the block numbers, the geometry from this layer doesn't need to be shown. You will, however, use the colors shown for platted and unplatted property so that anyone reading the map can identify the plat status from the color.

1 Turn on the Property Ownership layer. Right-click, and then click Properties. Set the symbol to No Color with an outline width of 0 for each of the three categories, and then close Properties. Note that unplatted property is blue and the platted property is purple. These colors will be used for the labels. The property with a pending plat will be labeled in orange.

2 Open the Label Manager dialog box and select the check box for the Base Map Group: Property Ownership layer. Click the layer name. Notice that there are suggested label classes based on the symbol categories. These classes will allow you to define separate rules and symbol types for each one. Clear the <all other values> check box and click Add. When prompted, click Yes to overwrite existing labeling classes.

8–1

8–2

3 You will now see these classes listed under the Base Map Group: Property Ownership layer.

4 Click the Plat Pending class name. Set Text String Label Field to Lot Number and the text color to Mango. In the Placement Properties drop-down list, click Land Parcel Placement. This will invoke preset rules that will provide the best placement for land parcels.

5 Repeat the process, selecting the color Heliotrope for the Platted Property class, and Yogo Blue for the Unplatted Property class. Set both with a Text String Label Field of Lot Number and use the Land Parcel Placement mode. Click OK to close the Label Manager and see the results. If you like, click View Unplaced Labels and notice that only a few labels at the extreme edges of the map went unplaced.

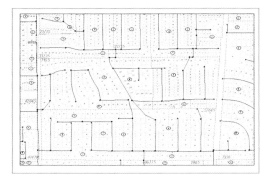

Label the sewer lines

The sewer lines are already symbolized and need to be labeled with their material and pipe size. The label will be on two lines, with the material shown over the pipe size. The label should also be centered on the line.

1 Click Label Manager and select the Sewer Data Group: sewerlin check box. Click the class name and notice that there are some predefined classes that you could use to match the symbol categories. The lines will all be labeled in one class, so you can ignore this suggestion.

2 Click Default and note the settings. In the Text String area, click Expression. Here you will build the expression for the label. It needs to contain the values from two fields: MATERIAL and PSIZE. The field MATERIAL has already been added to the expression, so select PSIZE and click Append to add it. Note that some extra code has been placed between the values. This will append a space between the values.

3 This text expression would put both values on the same line, and you wanted to put MATERIAL over PSIZE. To do this, replace the double quotes with **VBNewline**. Click Verify to preview the results. A feature is selected at random for the preview, so your display may show a different material type or pipe size. Click OK to close the Verification box.

4 Both fields are shown in the correct orientation, but the material name is spelled out. This will take up a lot of room on the map. To change it to the abbreviated codes for the Subtype values, clear the Display coded value description check box and verify again. Click OK to close the Verification box, and then OK again to accept the results.

8–1

8–2

5 In the Placement Properties area, click Properties. Click the Label Position tab and click Position, and then select Centered Curved. Click OK.

6 Select the Repeat label check box, and then click Interval. Set the minimum repetition interval to **800** map units. This will put a label every 800 feet along the pipes to ensure that there is no ambiguity in the map. Click OK.

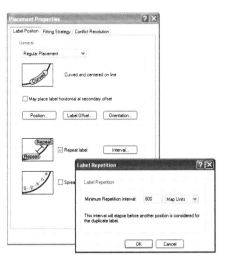

7 Click the Fitting Strategy tab and change the minimum feature size for labeling to **200** map units. This will keep Maplex from trying to crowd labels onto features that are less than 200 feet long.

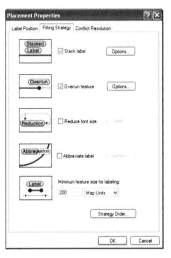

8   Click OK to close Placement Properties, and then OK again to close the Label Manager. Once you click OK, the changes are applied, so examine the results. Notice that the more layers you set up for labeling, the longer it takes to redraw the map.

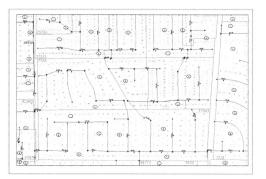

Label the manholes

The manholes have an attribute that notes how deep the flowline is from the surface. This lets field crews know how deep they can expect to dig in order to unearth a line for repairs and determine what type of equipment might be necessary.

1   The manholes are symbolized in the Sewer Data Group: sewernod layer. Click Label Manager and select the check box for this layer. Click the label class, and you are presented with three possible label classes for this layer. Clear all but manhole and click Add. Click Yes to overwrite existing labeling classes.

2   Click the new label class for Manholes to view its settings. In the Text String area, click Expression. Delete the existing expression and replace it with the following:

"D=" & Round([DEPTH],1)

This will round the depth value to one place past the decimal. For more information on ways to manipulate the expression values, click Help in the Expression dialog box. Click OK.

8-1

8-2

3 There might also be a problem with manholes for which the depth is not known. These have a value of zero and would clutter up the map while providing no useful information. Click SQL Query. A query to select only manholes exists. Add a statement to the query to also exclude values not equal to zero by typing the following at the end of the query statement:

> **AND [DEPTH] <> 0**

Click OK to close the query dialog box.

4 Accept the default placement positions by clicking OK and closing the Label Manager. Examine the results.

The labeling on the map is beginning to look rather complex, and the overall style is very nice. The final thing this map needs is street names.

Label the streets

Maplex has specific preset rules for street placement, much like the Land Parcel Placement you used earlier. The existence of sewer lines in the streets might interrupt the preset rules and require you to test other placement strategies to achieve the desired results. There are three classes of labels to make for the streets, and these will be based on the Class attribute. The first will be smaller text for residential streets (classes 6 and 7), a larger text for collector streets (classes 3 and 4), and the largest text for freeways (classes 1 and 2).

1   Turn on the Street_Centerlines layer and set its symbology to no color and a width of 0.

2   Open the Label Manager dialog box and click the Street_Centerlines layer. Notice that there are no label classes to select from, because there are no symbology classes that define them. Select the check box for this layer to turn on labeling.

3   You will need three label classes for this layer, and since none exists, you will make them yourself. In the Add label class text box, enter the name of the first class as **Residential**. Click Add.

4   Right-click the label class Default and select Delete Class. Confirm the deletion by clicking Yes.

5   Click the Residential class. Click SQL Query and build the query **[CLASS] = 6 OR [CLASS] = 7**. Click OK. This will limit the residential class to the codes for residential streets.

6 Set Text String Label Field to STNAME2. Set the font to Arial Narrow, bold, size to 10.

7 In the Placement Properties drop-down list, click Street Placement. This will set the placement parameters to a set of rules that is optimized for street names.

8 Click Position and select Centered Curved. Click OK.

9 Click OK to close the Label Manager and see the results.

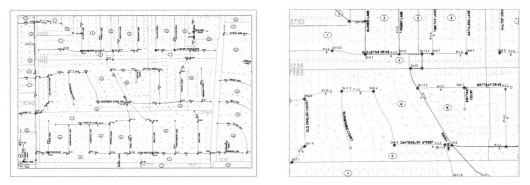

Build an abbreviation dictionary

This looks pretty good, but the full street suffix names are used. It might save space and make the map easier to read if the suffixes were abbreviated. To do this, you need to build and apply an abbreviation dictionary.

1 On the Labeling toolbar, click the Labeling drop-down arrow and select Abbreviation Dictionaries.

2 Click the Options drop-down arrow, and then click Import from Database Table.

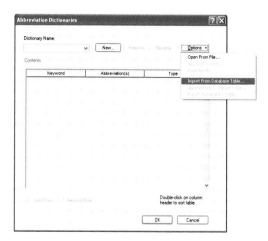

3 Browse to the Data folder and click Suffix.txt. Then click Add.

4 The suffixes from the U.S. Postal Service are added to the dictionary. Browse through the list to see how extensive it is. Click OK to store the dictionary.

5 Open the Label Manager dialog box and click the Residential class again. Click Properties, and then click the Fitting Strategy tab. Select the Abbreviate label check box and click Options. In the Dictionary Name drop-down list, click Suffix.txt, and then click OK. Click OK two more times to close Placement Properties and the Label Manager. The results now show the street suffixes abbreviated. The map looks a little cleaner with more room to place labels for all the features. If all the

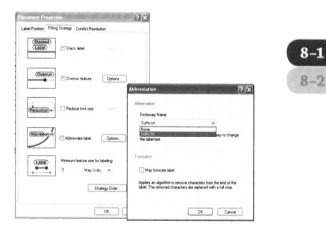

8–1

8–2

street names didn't show up, you could try some of the other Maplex placement options such as reducing the font size or compressing the letters together. It might also help to change the strategy order for the placement rules.

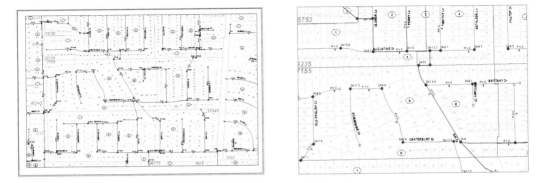

YOUR TURN

Add two more label classes named **Collectors** and **Freeways**.

Set up queries for the other two classes. The Collectors should contain classes 3 and 4, and the Freeways should contain classes 1 and 2.

Set the font to Arial Narrow, the size for Collectors to 14, and the size for Freeways to 17.

Set the placement mode to Street Placement and the position to Centered Curved.

Set the abbreviation dictionary to Suffix.txt.

Note: You'll find that the new label classes take on the setting of the existing class, so you will only need to change the SQL Query and the font size to complete each class.

The final results achieve a complex labeling strategy and take a fair amount of computer power to place all the labels. The trade-off is the amount of time it would take to place all the labels manually or to place them on several maps.

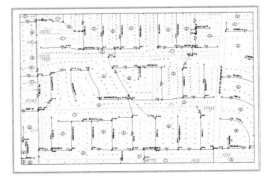

6 Save your map document as **GIST3\MyAnswers\Tutorial 8-1.mxd**. If you are not continuing to the next exercise, exit ArcMap.

Exercise 8–1

This tutorial demonstrated several techniques for using Maplex to dynamically place all the labels on a map.

In this exercise, you will repeat the process to create a similar map for the City of South Oleander Hills.

- Start ArcMap and open Tutorial 8–1e.mxd.

- Set up the options for using Maplex.

- For each set of labels, investigate the placement properties to achieve the best results.

- Label the Plat_Index layer with the SubID field. Position the label to the corner of the polygons using the font Arial, bold italic, size 20, in Seville Orange.

- Label the blocks using the Block Designations field. Place this in the special block number symbol.

- Label the Property Ownership layer with the Lot Number field. Use three classes with platted property in purple, unplatted property in blue, and plat pending property in orange.

- Label the sewerlin layer with the fields MATERIAL and PSIZE.

- Label the sewernod layer with the DEPTH field.

- Label the street_centerlines layer with the STNAME2 field. Use three classes with residential streets for class values 6 and 7, at size 10; collector streets for class values 3 and 4, at size 14; and freeways for class values 1 and 2, at size 17. Use the font Arial Narrow, bold.

- Build and apply an abbreviation dictionary for the street suffixes.

- Save your results as **GIST3\MyAnswers\Exercise 8-1.mxd**.

8–1

8–2

WHAT TO TURN IN

If you are working in a classroom setting with an instructor, you may be required to submit the map you created in tutorial 8–1.

Printed 11-by-17-inch map or screen capture of

        Tutorial 8–1

        Exercise 8–1

Tutorial 8-1 review

Labels are text that is placed automatically by the software using a feature attribute. With the use of Maplex, labels can be controlled very carefully. Once rules are defined, they are applied to all the labels. In most situations, the labels will be placed in the appropriate place, but not always.

Understanding the rules will help you to set rules to place as many labels as possible. All unplaced labels are shown to the user so that the rules can be modified to allow more labels to be placed.

STUDY QUESTIONS

1. Name three advantages of using dynamically placed labels.
2. Name three drawbacks of using dynamically placed labels.

Other real-world examples

One of the most popular uses of Maplex is labeling streets. A special set of rules is predefined for street names making street maps easier to create. Multipage map books benefit from dynamically placed labels where each page has the best chance of having each street labeled.

Another preset rule for dynamic labels is contour lines so that the labels appear aligned on a defined portion of the map making them easy to find. It can be problematic, for example, to have to search for an elevation on a contour map. By placing the labels in a consistent place, it is easier to interpret the map.

Tutorial 8-2

Adding geodatabase annotation

The geodatabase allows for the storage of user-placed annotation for use on multiple map documents. The annotation can be edited for unique situations, and linked to feature attributes to handle instances where the underlying data is changed. Annotations can be imported from other files, created by the user, or converted from dynamic labels.

Learning objectives

- *Convert labels to annotation*
- *Build new annotation*
- *Alter annotation symbology*
- *Build feature-linked annotation*

Introduction

The dynamic labeling used in tutorial 8-1 placed labels according to a set of rules. All the labels had to fit the rules, or else they were not added to the map. Another drawback with dynamic labeling is that the labels are stored with a map document and cannot be shared among several maps.

Annotations, on the other hand, are not created by a set of rules. They can be created one by one to reflect any style necessary. Because of this, annotation that would not normally be placed by dynamic labeling can be added and edited to fit special circumstances. The parameters for each piece of annotation are stored with the annotation, so customizing one piece will not alter the rest of the annotation.

Dynamic labels, such as those produced with Maplex, can appear in different places every time a map is zoomed or panned. However, this takes a lot of time for the labels to regenerate each time. Annotations are set in a fixed place and will not move unless edited by the user. This speeds up drawing time and ensures that the annotations will not move in conflict with other features or cause other labels to no longer draw. The trade-off for the increased speed is that the annotations may only be suitable for a small range of map scales.

8-1

8-2

Annotations are stored as an annotation feature class in the geodatabase. Once the user has completed a set of annotation, it can be reused over and over on as many map documents as desired. Annotations can also be linked to feature attributes, either when they are first created or when they are edited. In the event that the attribute of a feature changes, such as the owner of a house, the annotation will automatically update.

Annotations can be placed one by one, but a more efficient method is to use a conversion process. Dynamic labels can be set up to accommodate as many labels as possible, and then both the placed and unplaced labels can be converted to annotation. The unplaced annotations can be edited to a best-fit situation and stored. This can result in 100 percent of the annotations being placed.

The scale and complexity of the map plays a large part in whether to use dynamic labeling or annotation. If the map is in a relatively large scale and there are not a lot of features to deal with, dynamic labeling will produce good results. Of course, the results are only good for that map document. If the map is of a smaller scale and not all the dynamic labels can be placed or if there is a lot of interference with complex features, annotations would be a better choice. The resulting annotations can also be used on many map documents.

Labeling with placed annotation

Scenario The city planner has asked for some 11-by-17-inch zoning exhibit maps that will cover only a small part of Oleander. The City Council will be investigating future development sites and will use the zoning exhibits in its deliberations.

The map should display property boundaries but no annotation associated with the lots. It will have street names and zoning districts. The districts should be identified by classification with a text item.

Data The city planner made a prototype map with dynamic labels. The rules the planner set up put dynamic labels on a lot of the streets and zoning districts, but the map is not complete and doesn't look very good. The page size of 11 by 17 inches and the reference scale of 1 inch = 600 feet are already set. You will take the city planner's map and make annotations from the labels. Then you will clean up the map to make a better presentation.

Tools used ArcMap:

Convert Labels to Annotation
Editor toolbar
Edit Annotation tool
Unplaced Annotation tool
Construct Annotation tool
Masking layers

The Street_Centerlines layer has already been set up with dynamic labeling in place. Maplex and many of its settings were used to get the best placement of street names. However, many of the streets are not labeled because the rules could not handle every situation. You will convert these unplaced labels to unplaced annotation and manually fix the areas where street names were not labeled.

Convert dynamic labels to annotation

1 Start ArcMap and open Tutorial 8-2.mxd.

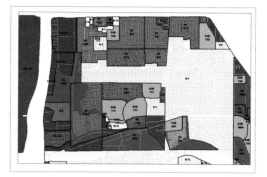

2 Right-click ZoningDistricts and select Convert Labels to Annotation.

3 Notice at the top of the dialog box that the annotation is set to be stored in a database rather than the map document. This will make the annotation available for use in other maps and for other users. The reference scale is also set. Change the Create Annotation For value to Features in current extent. Since this is not intended to be a map of the entire city; it would waste a lot of time and space to convert all the labels.

4 Note that the new annotation will be linked to the attributes of the features that were used for the dynamic labels. Change the Annotation Feature Class output file name to **ZoningDistrictsAnno7200** to give future users an idea of the scale for which this annotation is appropriate. Then click the Parameters button next to the file name.

5 Note the settings for Feature-Linked Editing Behavior. If new features are added or existing features are modified, the annotation will be automatically updated. Accept the defaults for these settings and click OK.

6 Finally, make sure that the "Convert unplaced labels to unplaced annotation" check box is selected. Notice that the annotation will be stored in the City of Oleander geodatabase. Click Convert. The map looks exactly the same as before, but an important change has taken place. The dynamic labels for the ZoningDistricts layer have been turned off, and the new annotation layer has been added and made visible. If you were to zoom and pan the map, the zoning annotation would not reposition itself to fit the map scale, and the map would also refresh faster. If you try this, be sure to zoom to the full extent before continuing.

YOUR TURN

Turn on the Street_Centerlines layer and convert its labels to feature-linked annotation. As with the zoning districts, add the reference scale to the end of the output file name. Be sure to convert only the features in the current map extent.

In the table of contents, you'll see the new annotation layer. Note that the label classes you defined earlier are now subtypes in the annotation layer.

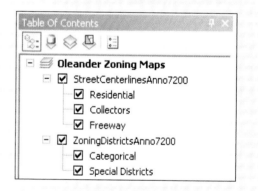

7 Right-click the Special Districts annotation layer and click Properties. Make sure you are on the General tab. Here, you will see the settings for scale dependency and a description. Set the minimum scale to 1:7,200 and add a description of **Annotation for special zoning districts.** Click Apply.

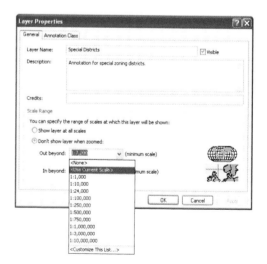

8-1

8-2

8 Now click the Annotation Class tab. This will display all the parameters of the dynamic labels: the font information, the queries, and the placement options. These cannot be changed but appear only as a reference. Any new features that are created will use these parameters for annotation placement. Click OK to close the dialog box.

YOUR TURN

Set a minimum scale of 1:7,200 for the Categorical zoning annotation. Also, give it a description of **Annotation for standard zoning categories**.

Setting these rules for the zoning annotation will help control when the labels are displayed as well as provide important information to others who may use these labels in the future.

Edit the street name data

There are many problems with the street names. Maplex wasn't able to place all the street names, and in some areas, it placed several instances of the same name. Even some of the placed annotations can be made a little better.

1 Turn off the ZoningDistrictsAnno7200 layer. Make sure the Editor toolbar is visible and start an edit session. Make Street_CenterlinesAnno7200 the only selectable layer. (**Hint:** Right-click the layer, point to Selection, and click Make This the Only Selectable Layer.)

2 Load and zoom to the Street 1 bookmark. You will need to be able to see the street centerline while editing the annotation, but not at the full map scale. Set the symbology for Street_Centerlines to the Residential Street symbol. Then set a display threshold to not display the layer if you zoom out beyond 1:6,000.

Layer Properties

General | Source | Selection | Display | Symbology | Fields | Definition Query | Labels | Joins & Relates | Time | HTML Popup

Layer Name: Street_Centerlines   ☑ Visible

Description:

Credits:

Scale Range

You can specify the range of scales at which this layer will be shown:

○ Show layer at all scales

◉ Don't show layer when zoomed:

Out beyond: 1:6,000 ▾ (minimum scale)

In beyond: <None> ▾ (maximum scale)

OK   Cancel   Apply

3 The name State Highway 121 appears too many times on the map. Click the Edit Annotation tool on the Editor toolbar, press and hold the Shift key, and select the freeway names shown in the graphic.

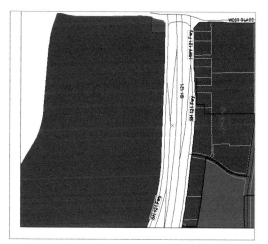

Editor

Editor ▾

4 On the Editor toolbar, click the Attributes button. This will open the Attributes dialog box for the selected features.

Editor

Editor ▾

Attributes

8-1

8-2

5 You can see in the Attributes dialog box that all the settings for the features are stored here. Since these three are extras, highlight all three and click Delete.

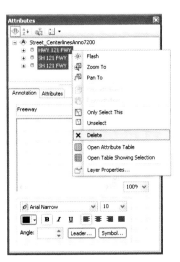

6 Select the remaining SH 121 annotation shown on the map. Update the font size in the Attributes dialog box to **15** and move the annotation farther south and centered on the freeway. If your text flips during the process, click the Attributes tab and change the value of FlipAngle to 80.

7 Zoom back to the full extent. Select and delete the other two annotations for SH 121.

YOUR TURN

Load and zoom to the Street 2 bookmark. Clean up the annotation for Hwy 360 in the same manner as SH 121, making the font size **15** and moving the label to a better position.

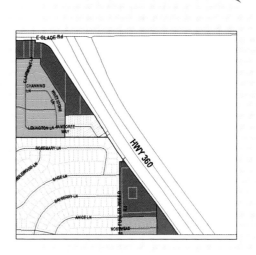

Place the unplaced annotation

There are several areas where Maplex was unable to place labels for the street centerlines. As annotations, these can be placed manually from the Unplaced Annotation list.

1 Use Customize Mode to add the Unplaced Annotation button [icon] to the Editor toolbar.

2 Load and zoom to the Street 3 bookmark. Click the Unplaced Annotation button on the Editor toolbar. Resize the window and drag it to a location where both the window and the map area are visible.

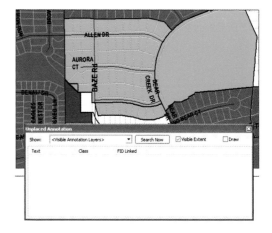

8-1

8-2

3 In the Unplaced Annotation dialog box, in the Show drop-down list, click Residential, and then select the Visible Extent and Draw check boxes. This step will find unplaced annotations for the Residential layer only, in the current drawing extent, and draw them on the map temporarily. Click Search Now. The unplaced annotation is drawn with a red box around it. The suggested places are not very good, but once added as regular annotation, they can be altered to fit. For each annotation in the search list, you have the option of adding it to the map or deleting it. A few of these annotations can be added and altered, and a few need to be deleted.

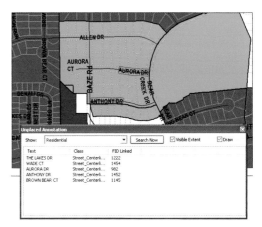

4 Click AURORA DR on the Unplaced Annotation list and watch it flash on the map. To place it, either right-click the street name on the list, and then click Place Annotation, or use the shortcut and just press the spacebar. The red box around the annotation disappears and the annotation becomes part of the map.

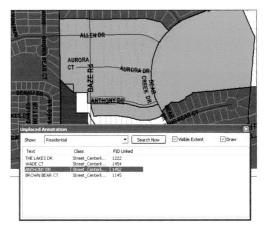

5 The annotation needs to move a little to the left to be centered. Select it with the Edit Annotation Tool and move it. You will notice that although it was created with a curve to follow the line, it is no longer following the line's curve. Set this up by right-clicking the annotation. Then point to Follow and click Follow Feature Options.

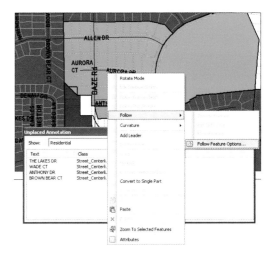

6 In the resulting dialog box, change the Make Annotation value to Curved, and set Constrain Placement to On the line. Click OK.

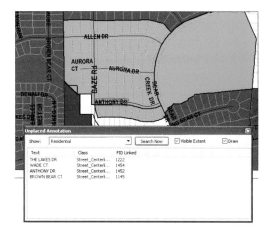

7 The annotation for AURORA DR is currently multipart, as witnessed by the two boxes surrounding it. In order for it to follow the street feature, it has to be a single part. Right-click the annotation and select Convert to Single Part.

8 Now slide the annotation to the left until it is centered on the feature. Notice that it curves to follow the feature.

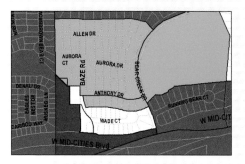

YOUR TURN

Try placing the annotation for ANTHONY DR and WADE CT. Move them into position, following the underlying features. Then correct the placement for RUNNING BEAR CT. Also, move BEAR CREEK DR into a better position, and delete the extra annotation that was made for this street.

8-1

8-2

There are still a few unplaced annotations to deal with. The remaining annotations will take some special editing to make them acceptable.

9 Click LITTLE BEAR TR in the Unplaced Annotation window. You'll see that there is a duplicate of an existing annotation. Highlight both lines that show Little Bear Trail, right-click, and then click Delete.

10 Select and place BROWN BEAR CT. Since it is at the end of BROWN BEAR WAY, it is acceptable to label it simply as CT. Open the Attributes dialog box and change the text to **CT**. Then move the text into the wide part of the cul-de-sac.

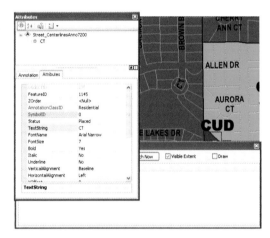

11 Select and place THE LAKES DR. There is no way it will fit here. Try moving it. You'll find that it is tied to the feature. Right-click and clear the Follow Feature Mode check box. Now move it up above DENALI DR.

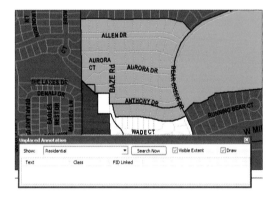

12 In order for the annotation to have some relationship to its feature, right-click it, and then click Add Leader.

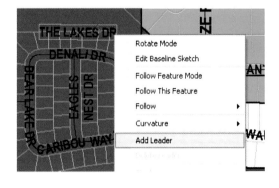

13 Clear the Draw check box and close the Unplaced Annotation window. Stop the edit session and save your edits. The results look pretty good. There are other areas of the map that could use this type of attention. It would be time-consuming, but the results are worth it.

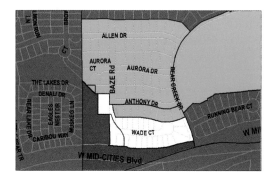

Add new annotation

1 Load and zoom to the Street 4 bookmark. These street names were not placed by Maplex because of a coding error. Once the coding error is fixed, the street names can be added. Start an edit session and set Street_centerlines as the only selectable layer. Select the five street segments in this map extent that do not have names.

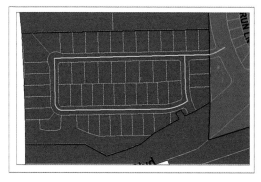

2 Open the Attributes dialog box and change the class for all five lines to **Residentials**. Close the Attributes dialog box and save your edits.

3 Verify that Street_Centerlines and Street_CenterlinesAnno7200 are the only selectable layers.

4 Click the Street_CenterlinesAnno7200: Residential template, go to Construction Tools, and select Follow Feature.

5 Move the tool crosshair over the street centerline at the top of the subdivision and press Ctrl + W. This will get the street name from the Attributes table of the centerline.

6 Put the cursor over the street centerline and click. The annotation will lock to the feature. Move the annotation into position and click again to place it.

YOUR TURN

Using the construction tools, place names on BLACK BEAR DR and LOST VALLEY DR. When you are done, stop the edit session and save your edits.

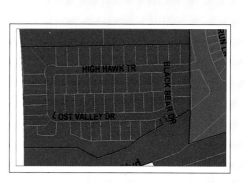

Edit the zoning annotation

Each of the zoning categories has annotation that displays its code. Most of these annotations are fine, but a few need minor touch-ups. With dynamic labels, you would not be able to change the parameters of a single piece of text, but with geodatabase annotation, you can.

1 Start an edit session. Turn on the ZoningDistrictsAnno7200 layer, and make it the only selectable layer. Load and zoom to the Street 5 bookmark.

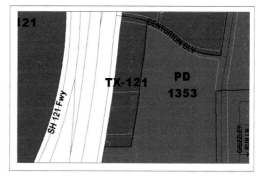

2 The annotation for TX-121 needs to rotate to align with the zoning district boundary. Use the Edit Annotation tool and select the text. (**Hint:** Make sure you are selecting the text with the Annotation Edit tool, and not the regular Selection tool.) A box is drawn around the selected feature, but there are some aspects of the box that are different from a normal selection box. In the center is a small X. This is the lock-down point of the text. To move the X, pause the pointer over it, and press and hold the Ctrl key. Then move the X to a new location. At the top is a red triangle. Pressing and dragging this triangle will make the text

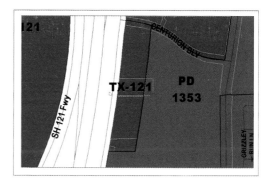

8-1

8-2

larger or smaller. At the two bottom corners are blue arcs. Clicking and moving one of the arcs will cause the text to rotate about the opposite corner. Clicking the text anywhere else will result in a standard move. The cursor will change to indicate the action being taken.

3 Click one of the blue arcs in the selection box and rotate the text to match its surroundings. Then move it into position.

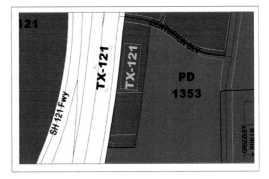

4 Load and zoom to the Street 6 bookmark. The label for this large R-1 area is too small for the size polygon it represents. Select the label and use the red triangle to make it larger. Then move it to a better position. Save your edits before moving on.

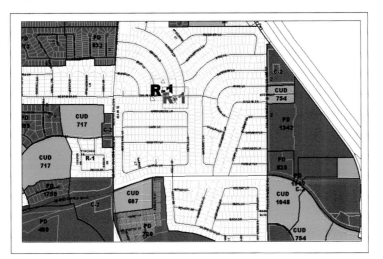

Mask the background layers

1   Load and zoom to the Street 7 bookmark. The city planner would like the zoning codes to be more legible and not be obstructed by the parcel lines, street edges, or zoning district boundaries. The planner tried using text halos, but the results looked bad and in some cases obscured entire zoning districts.

There is a better way to make the codes legible, by using layer masking. Layer masking involves making an outline of the features to be masked, and then using the outline with a special mask setting in ArcMap. For this tutorial, the masks for the zoning annotation are provided, but users with an ArcInfo license could make their own outlines using the Feature Outline Masks tool.

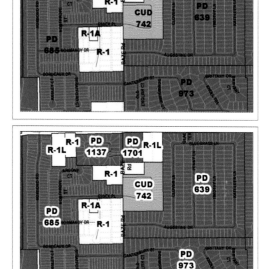

2   Turn on the ZoningDistrictsMasks layer in your table of contents. This looks worse than the text halos.

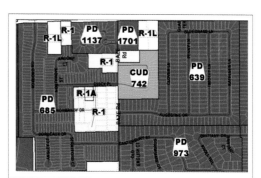

8-1

8-2

3   Turn off the ZoningDistrictsMasks layer. Right-click Oleander Zoning Maps in the table of contents, and then click Advanced Drawing Options.

4 Select the Draw using masking options specified below check box. In the Masking Layers pane, select Zoning Districts Masks. In the Masked Layers pane, select the ZoningBoundaries and Lot Boundaries check boxes. Then select the check box for Enable to associate levels to masked layers. Click OK to close the dialog box. The area of the polygons created by the masking tool is cutting out the lot lines and zoning district boundaries. But instead of leaving an ugly white blob, the background zoning district symbology shows through.

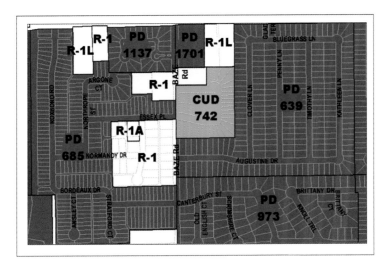

5 Save your map document as **\GIST3\MyAnswers\Tutorial 8-2.mxd**. If you are not continuing to the next exercise, exit ArcMap.

Exercise 8–2

This tutorial showed many aspects of working with annotation, from altering placed annotation to creating masks for annotation.

In this exercise, you will complete some of the areas that were not handled with Maplex.

- Continue working with Tutorial 8–2.mxd.

- Load and zoom to the Street 8 bookmark.

- Look for any unplaced annotation and place it.

- Correct any problems you might find with existing placed annotation.

- Load and zoom to the Street 9 bookmark.

- Correct the errors with placed or unplaced annotation at this location.

- **Note:** If you move or resize any zoning district annotation, you can manually edit its mask in the ZoningDistrictsMasks layer.

- If you have an ArcInfo license available, create a mask layer and have it mask the lot lines around the street names. (**Hint:** You will need to choose an Input Layer, Output Feature Class, and Calculation Coordinate System.)

- Save your results as **\GIST3\MyAnswers\Exercise 8–2.mxd**.

WHAT TO TURN IN

If you are working in a classroom setting with an instructor, you may be required to submit the finished map you created in tutorial 8–2.

Printed 11-by-17-inch map or screen capture of

Tutorial 8–2

Exercise 8–2

Tutorial 8-2 review

Annotations are used when the text being placed needs to stay in a set location. Annotations will not move when the map is zoomed or panned as labels do. There is, however, a large overhead in terms of time in placing a lot of annotation.

One of the best ways to create annotation is to use dynamic labeling to place as much text as possible. Then the labels can be converted to annotation and the unplaced items can be located in the proper place. Annotation can also be created manually. Several shortcuts make it easy to grab text from a feature's attributes and use it as annotation for that feature.

Annotation can also be linked to the attributes table of a feature. If the values used for the annotation change, the text shown on the map will also change. This is useful for text such as the owner's name on property or other features where the value changes on a regular basis.

STUDY QUESTIONS

1. What is the difference between labels and annotation?

2. Name three benefits of using annotation over labeling.

3. Name three drawbacks of using annotation over labeling.

Other real-world examples

Maps with background text, such as state names or district names, often use annotation to name these features. The annotation is set as background text and will not move or change size as the map is zoomed or panned. The annotation can also be set to have other features draw on top of it.

Annotation is often used for naming features with a static location such as buildings. The names are placed in relation to the feature to give the best presentation on the map. These names can be altered individually for specific fonts and sizes.

References for further study

There are a number of additional resources to help with labeling and annotation. **ArcGIS Desktop Help** lets you use keywords to search for a title or topic. You can access ArcGIS Desktop Help in ArcMap or ArcCatalog by clicking Help on the main menu. Then click the Search tab, type a keyword, and click Ask. Use the keywords provided here to search for additional learning resources on many of the concepts taught in this chapter.

ESRI has developed a number of online courses on a wide variety of pertinent GIS topics. **ESRI Self-Study (Virtual Campus) courses** are an excellent resource for students that will supplement information covered in this course. ESRI Virtual Campus courses can be accessed at `http://training.esri.com/gateway/index.cfm`.

ArcGIS Desktop Help search keywords

labels, annotation, text, dynamic labels, feature-linked annotation, Editor toolbar, Edit Annotaion tool, annotation feature class, Maplex

ESRI Self-Study (Virtual Campus) courses

The following ESRI Virtual Campus classes may be helpful:

1. Creating and Editing Labels and Annotation
2. What's New in ArcGIS 9 Labeling and Annotation

9

Exploring cartographic techniques

Tutorial 9-1

Building cartographic representations

Each type of data created in a GIS framework is one of three major types: points, lines, or polygons. Symbolizing raw data is quite limited, making it difficult to produce a digital map that has the rich symbology of a hand-drawn map. The addition of cartographic representations to ArcGIS is the solution to this problem. A new set of features is created in association with the point, line, or polygon features that are stored in a GIS dataset. These new features exist solely for the purpose of building a rich representation of the data, which results in a more cartographically pleasing map.

Learning objectives

- *Build cartographic representations*
- *Establish representation rules*
- *Add advanced symbol effects*

Introduction

Cartographic representations are used to build a complex symbology for standard GIS features. Multiple symbol layers and effects are created and manipulated to produce an almost unlimited variety of displays. The standard symbology in ArcMap, on the other hand, uses TrueType fonts and glyphs to create the symbols. This limits both the complexity and color palette that can be used for symbols. Overcoming this limitation is one of the strongest reasons to use representations over the standard symbology tools.

Representations can be created in several ways. The most common is to convert the symbology for existing layer classifications. It is also possible to create representations from layer files or to create empty representations and build the symbology manually. However they are created, they are stored in separate files yet maintain an association with the source data. An additional benefit of using representations is that the shape of the representation can be altered without changing the underlying data. For instance, a road and rail line may be so close together that standard symbols would cause them to appear to overlap. With a representation, the lines can be separated for clarity without disturbing what may be very accurate source data.

There are three basic types of symbols used for representations: markers, strokes, and fills. These can be combined with a series of geometric effects, such as dashes or offsets. To make a representation symbol, one or more of the symbol types are added to a display list, and then various effects such as dashes or gradient fills can be built for them. Finally, all the symbols are layered together to create the display. In addition to the effects, there are also rules that can be set to handle how representations start, stop, or interact with other representations. For example, a dashed line can be set to force a half symbol at both the start and stop of the line, in contrast to the regular symbol tools that leave this to chance. Representations can also detect their interference with other representations. The features may have plenty of separation and look fine when they are not symbolized, but once symbols are added, the symbols used may touch or overlap. A special representation tool can detect conflicts and allow you to fix them, making all the symbols visible. The result is symbology that is closer to the way you would draw it if you were able to draw each piece individually.

The representations are linked to their features but may be altered individually for special circumstances. For example, representation rules can be set for the lakes layer and be applied to all the lakes. Then a single lake might be altered to highlight it for study without affecting the way other lakes are displayed.

At any time, users can return to the source data for updates and edits. The new features will be added to the representations and be ready for inclusion on any map. Working with these two sets of data will allow for very accurate data maintenance while still providing a cartographically pleasing map.

Building new symbology

Scenario   The City of Oleander had new planimetric data collected this year and the city manager wants you to produce a wall map for the City Council chamber that will show off this data. You'll produce an 11-by-17-inch map as a proof of concept before committing the time to a large wall map.

The planimetric, or physical feature, data includes a wealth of features to work with, but for this map, you'll limit the investigation to street centerlines, fences and retaining walls, recreational areas, trails, tree mass, and hydrography.

Data   The following layers have simple symbology set up, which serves as the starting point for the representation:

Tree Inventory	Fences, retaining walls, etc.
Trails	Recreational Features
Street Centerlines	Bodies of Water

9–1

9–2

Tools used   ArcMap:

Convert Symbology to Representation
Representation Marker Selector
Graphic Quality: Detect Graphic Conflict
Representations: Erase tool
Representations: Direct Select

Create cartographic representations

The first step in making representations is to convert the symbology from the existing files. The new file names will have "_Rep" appended to the end to remind you that the file contains representations. For this tutorial, you will make a new group layer to contain the representations.

1 Start ArcMap and open Tutorial 9–1.mxd. A sample area is shown for testing the representations.

2 In the table of contents, right-click the Physical Features data frame. Click New Group Layer. Rename the group layer **Representation Group**.

3 Right-click the Trails layer, and then click Convert Symbology to Representation. **Note:** Make sure that you do not have a Search or Catalog window open during this step as it will cause areas to not be converted to representations.

4 The resulting dialog box has a lot of information about the conversion. The first is the name of the output file. The default is to append "_Rep" to the end of the layer name. The next two inputs are for fields that contain the internal workings of the representations. Since these are managed by ArcMap, it is best to leave them alone. The next is to determine how to handle edits to the representations. It is possible to have the underlying data change when the representations are changed. Finally, there is the choice of converting the entire set of features to representations, or just those in the current map extent. For this trial, convert only the features in the current extent. When your dialog box matches the graphic shown here, click Convert to start the conversion.

5 Drag the new layer into the Representation Group in the table of contents.

YOUR TURN

Convert these remaining layers to representations:

> TreeInventory
>
> Street Centerlines
>
> Fences, retaining walls, etc.
>
> Recreational Features
>
> Bodies of Water

Remember to convert only the features in the current extent and to drag the results into the Representation Group.

9–1

9–2

6 When you have converted all the layers, minimize the Raw Data Group and turn the layers off. The map will look the same. To see where the representations are stored, right-click the Trails_Rep layer, and then click Properties. Click the Symbology tab. You'll notice a new header in the Show box for representations. Although the symbol looks the same, it is now converted to a style that is managed with more complex tools. Leave Layer Properties open for the next step. The center pane shows the current symbol rules that are added to the representation. The buttons at the bottom can be used to add more rules or set rule options. The right pane shows the symbol layers for the representation. Buttons below this pane can be used to add marker layers, stroke layers, or fill layers, as well as to set layer options.

Build a representation

A single line representation exists for the Trails_Rep layer. A more desirable symbol would be a tan line with a dashed gray line on top of it.

1 Click the Color box for Rule_1 and change it to Medium Sand.

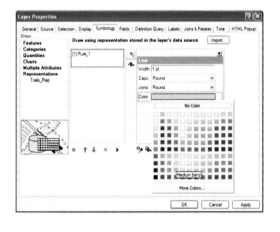

2 Below the Line pane, click the Add new stroke layer button to create a new layer.

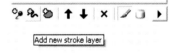

3 Change the color to Gray 40%.

4 Above and to the right of the Line pane, click the Plus Sign button to add a geometric effect. Click Dashes, and then click OK.

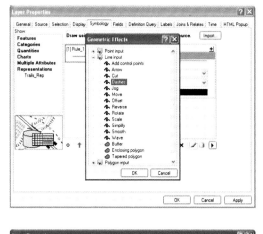

5 A new set of parameters is added to the layers area. This shows the pattern for the dashes and the rule to use for ending the symbol. Click the drop-down arrow to see the other choices, but accept the default value. A preview of the final symbol is shown on the left. Click OK.

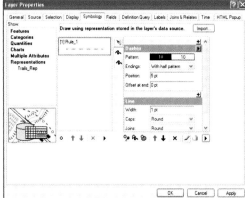

6 A warning will note that the changes will be saved with the feature class in the geodatabase. Select the Don't warn me again in this session check box and click OK. The new symbol is used to draw the trails on the map. If you don't like the way it looks, you can return to the Properties dialog box and alter the representation.

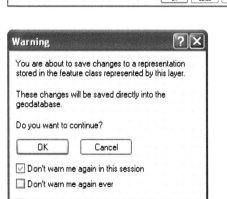

Create point representations of trees

9–1

9–2

Included in the table of contents is the TreeInventory_Oleander_Rep layer. This layer contains the GPS points for every tree in the parks and rights-of-way in Oleander. A field called Type was used to classify this layer, and the symbols for the classifications have now been converted to representations.

1 Load and zoom to the Bear Park bookmark. This shows the tree inventory points in Bear Park.

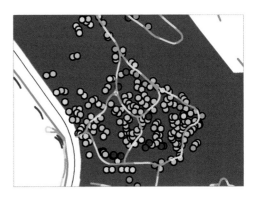

2 Open the properties of the TreeInventory_ Oleander_Rep layer. Make sure you are on the Symbology tab and that you have clicked Representations. You will see a rule for each of the 37 tree types. Each is symbolized by a single marker layer. Scroll down and select symbol number 13, the elm.

3 Click the dot marker in the Marker pane. This opens the Representation Marker Selector. Scroll through the list and note the complexity of the markers, which are much more elaborate than the TrueType markers available with standard symbology.

4 None of these markers would represent a tree very well, so you will need to build one. Click Properties to open the Marker Editor. Then click the symbol. In the right pane, you'll see the properties that make up this symbol. Press Delete to remove this symbol and clear the way to build a new one.

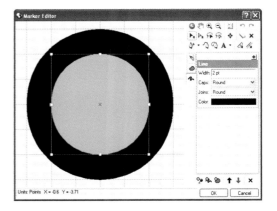

5 In the Marker Editor tool area, click the Create Glyph tool.

6 The Symbol display will open. Set Font to ESRI US Forestry 2. A large variety of tree symbols is displayed. Some trees are shown in profile view, and some are shown as if viewed from above. For Unicode, type **162** to select the tree symbol shown in the graphic. Click OK.

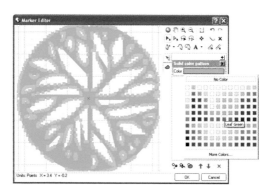

7 A new marker is created from this element in the Font table. Notice that the wireframe depiction also shows the line nodes that could be moved to customize the shape. Set the color to Leaf Green. Click OK to close the Marker Editor, and then OK again to close the Representation Marker Selector.

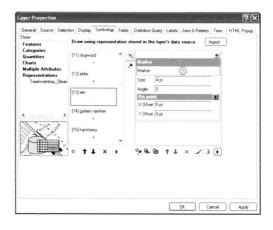

8 The tree symbol is shown as a representation in the Layer Properties dialog box and is ready to use. Leave the default font size at 4 pt.

9 Click Apply and move the Layer Properties dialog box to view the results. This looks much more like trees than the colored dots. There are thirty-six more tree types, and each type would need a unique symbol.

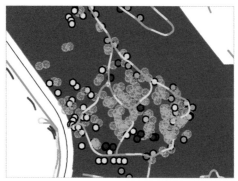

YOUR TURN

Create new tree representation symbols for these types:

Text	Unicode value	Color	Font size
Hackberry (15)	169	Leather Brown	3.3
Pecan (26)	198	Dark Olivenite	5.2
Ash (1)	107	Peacock Green	6
Oak (23)	189	Dark Umber	7.2
Osage orange (24)	121	Raw Umber	5.8

Close Layer Properties and examine the results.

Even after this extra work, there are a lot more tree symbols to make. But you can start to see how complex and realistic the tree representations can be.

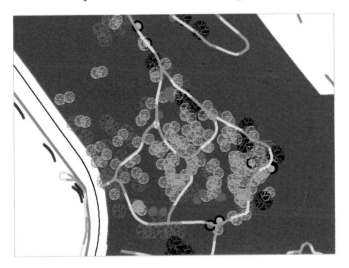

The tree symbols are being drawn in a seemingly random order. It would look nicer if the darker tree symbols drew first, and then the lighter colored symbols were layered on top.

10 Open Layer Properties again and confirm that you are in the Representations area of the Symbology tab. At the bottom of the Marker pane, click the Layer Options button ▶, and then click Symbol Levels.

11 By altering the order of the symbols in the list, you can alter the order in which they draw on the map. Select the "Draw this layer using the symbol levels specified below" check box. The dialog box becomes active. Scroll down and find elm and click the upper arrow to move it to the top of the list.

12 Continue moving symbols to the top of the list and put them in this order:

elm

pecan

ash

hackberry

osage orange

oak

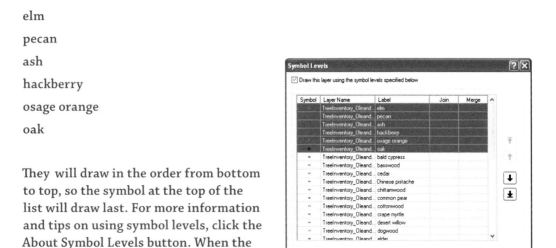

They will draw in the order from bottom to top, so the symbol at the top of the list will draw last. For more information and tips on using symbol levels, click the About Symbol Levels button. When the symbols are in the correct order, click OK to close Symbol Levels.

13 One more thing: There are way too many trees to appear in the legend. This can be restricted by using the Legend Options button. Click Layer Options, and then click Legend Options.

14 Remove all the items from the Legend Items pane with the << button. Then select the six that now have custom symbols and add them back to the Legend Items pane with the > button.

15 Finally, click OK in Layer Properties to see the results. With time, a very realistic depiction of the trees can be produced. The Representation tools are flexible enough to accommodate a much more complex symbology than the standard symbols can.

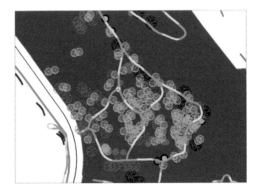

Create polygon representations of parks

The trees look great, but not against the purple park symbol. The symbology for the parks is based on the polygons in the Recreation_Rep layer. The layer contains polygons for the sports areas (parks) and swimming pools. Representations can be built to show the parks in a more realistic manner.

1 Open Layer Properties for the Recreation_Rep layer and go to the Representations area of the Symbology tab. You will see the default representations that were converted from the existing symbology. Click the Sports Area/Court name and change it to **Park**.

The representation for these polygons is a solid purple. It would be more realistic to make the parks representation look like grass, with light green colors.

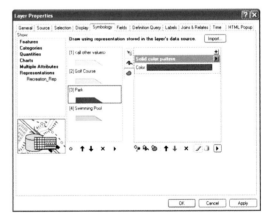

2 In the layers area, click the polygon layer and change the color to Tzavorite Green.

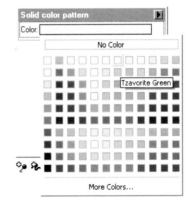

3 Next, click the Add New Marker Layer button at the bottom.

4 A new layer is added to the list, showing the default black square. Click the symbol to open the Representation Marker Selector, and then click Properties to open the Marker Editor.

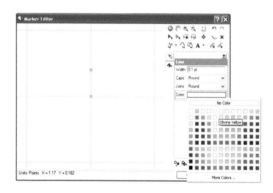

5 Click the black box symbol, and then press Delete. Next, select the Create Line tool from the Marker Editor tools.

6 Click in the center of the grid, move the pointer up a bit, and then double-click. This will create a new line. Set the color to Olivine Yellow.

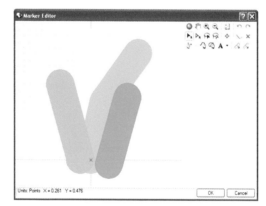

7 This will be the first line in what will become a tuft of grass. Draw three more lines, using the colors Light Apple, Macaw Green, and Leaf Green. The order isn't important because it's supposed to have a random, grassy feel.

9–1

9–2

8 Click OK to close the Marker Editor, and then OK again to close the Marker Selector.

9 Change the marker size to **1 pt**. Notice that the placement type is currently set to Polygon center. Click the right arrow, and then under Marker Placements, click Randomly inside polygon. Click OK. This will place the marker randomly many times within the polygon area.

10 Set the X step and the Y step values to **6 pt**, and then click OK.

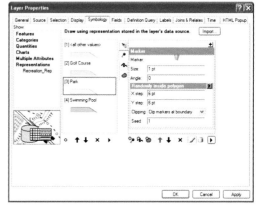

11 In the table of contents, move the Recreation_Rep layer to the bottom of the list. The new marker is spaced randomly inside the park area. While this representation may not look overly realistic, it is a little better than using the solid, purple color for grass. You can go back and tweak the colors, size, and offset spacing to try for a better look. In the meantime, this is a good example of using marker, stroke, and fill symbols in combination to create the representation rule.

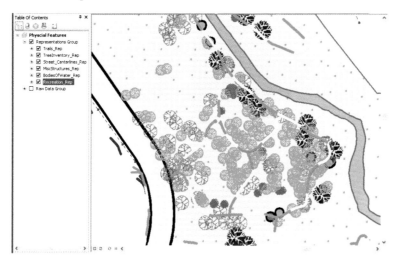

Create polygon representations of water

There are several lakes and streams in the Bear Park area, and these are currently symbolized with the standard fill color. A much better effect, however, can be created using representations. It would be nice to have all the lake and stream colors a gradient blue to give the water a feeling of depth.

1 Move the BodiesOfWater_Rep layer to just below Trail_Rep in the table of contents. Load and zoom to the Dog Park bookmark. This is an area to the west of Bear Park where pet owners can let their dogs run free. It includes a nice fishing pond, where you can test the representation rules.

2 Open Layer Properties and go to the Representations area of the Symbology tab. You can see that there is only one representation rule here for water and two representation layers—one for solid fill and one for the outline. You will modify this for a better-looking lake.

3 Click the Solid color pattern arrow, and then click Gradient.

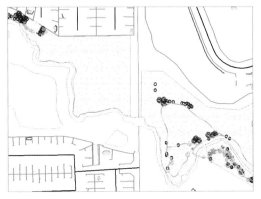

4 Set Color 1 to Sugilite Sky and Color 2 to Lapis Lazuli. Set Style to Circular and click Apply.

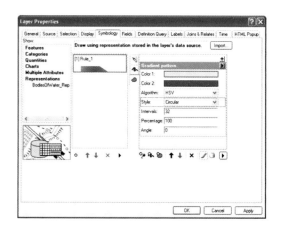

5 Click OK to see the results. This is a much more realistic view of the city's water bodies, giving them a dynamic, rippling effect.

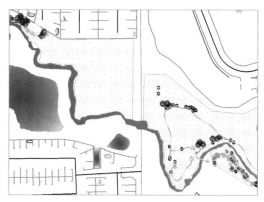

Create linear representations of fences

Next you will look at the fences in the MiscStructures_Rep layer. The thick green line being used for fences is not very realistic. A better symbol might be a thinner, dashed line with a dot every so often to represent fence posts.

1 Open Layer Properties for MiscStructures_Rep and make sure that you are in the Representations area of the Symbology tab. Click the representation rule for Fences and note that it is a single stroke layer.

7 To set up the representations for rule 4, create the following settings:

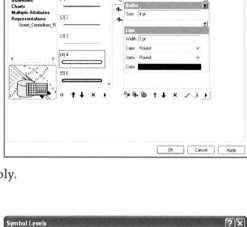

- On the existing stroke layer, set the width to **8 pt**, the Caps and Joins to **Round**, and the color to Gray 10%.

- Add a new stroke layer. Set its width to **2 pt** and the Caps and Joins to **Round**. Keep the default color set to Black.

- Add a geometric effect of buffer to the new line. Set its width to **4 pt**.

- When these settings are in place, click Apply.

8 Finally, open the Symbol Levels dialog box again. Select the Join and Merge check boxes for both Labels 4 and 6. This will merge the overlapping endpoints of the streets to create a clean corner. Click OK to close the dialog box.

9 Make sure that the stroke layer with the buffer effect is at the bottom of the layers list. Click OK to close Layer Properties, and then examine the results. The road and street representations are starting to look good. Don't worry about some of the other roads not drawing, however. They will be fixed as representation rules are made for them.

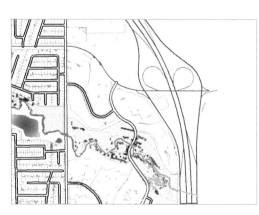

9–1

9–2

YOUR TURN

For this sample area, you will need to complete the representation rules for numbers 1, 2, and 15.

For rule 1 (label 1):

- On the existing stroke layer, set the width to **12 pt,** the Caps and Joins to **Round**, and the color to Gray 10%.

- Add a new stroke layer. Set its width to **2 pt** and the Caps and Joins to **Round**. Keep the default color set to Black.

- Add a geometric effect of buffer to the new line. Set its width to **6 pt**.

- When these settings are in place, click Apply.

For rule 2 (label 2):

- On the existing stroke layer, set the width to **10 pt**, the Caps and Joins to **Round**, and the color to Gray 10%.

- Add a new stroke layer. Set its width to **2 pt** and the Caps and Joins to **Round**. Keep the default color set to Black.

- Add a geometric effect of buffer to the new line. Set its width to **5 pt**.

- When these settings are in place, click Apply.

For rule 7 (label 15):

- On the existing stroke layer, set the width to **3.5 pt**, the Caps and Joins to **Round**, and the color to Gray 10%.

- Add a new stroke layer. Set its width to **2 pt** and the Caps and Joins to **Round**. Keep the default color set to Black.

- Add a geometric effect of buffer to the new line. Set its width to **1.75 pt**.

- When these settings are in place, click Apply.

Finally, update the Join and Merge settings in the Symbol Levels dialog box.

10 Click OK to close Layer Properties and see the results.

Turn off the Representation Group and turn on the Raw Data Group to see the difference between the maps. When you are done, turn off the Raw Data Group and turn back on the Representation Group. There is quite a difference.

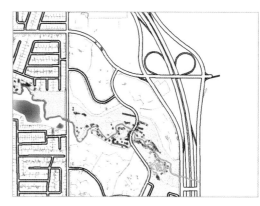

Edit the cartographic representations

Your new map looks very realistic and has a rich texture for a digitally created map. There are just a few minor problems with the rules, however. They create suitable representations for almost every feature, but there are a few things that need correcting. First, the river doesn't flow over the road as shown. Second, the lakes in Dog Park were not full when the planimetric data was collected, and now, they are larger.

1 Zoom to the Dog Park bookmark. Start an edit session and make the BodiesOfWater_Rep layer the only selectable layer. Add the Representation toolbar to your map document. This toolbar will let you modify an individual representation without modifying the rules for the rest of the features.

2 On the Representation toolbar, click the Select tool ▶ℝ and select the part of the river that goes over the road. Then click the Erase tool.

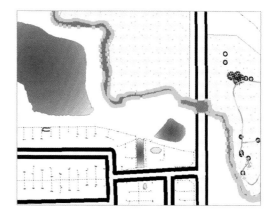

3 Point to the river and drag the eraser over the overlapping area. If your eraser is too big, right-click, and then click Options to set a more appropriate size.

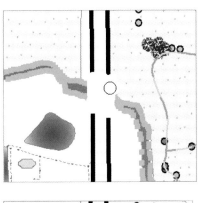

4 When you are done, release the mouse button, and when the drawing is refreshed, the overlap will be taken care of. If you accidentally erase too much, you can click the Undo button and try again.

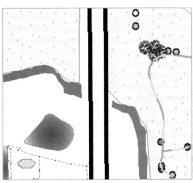

YOUR TURN

Pan to other areas in Bear Park and erase the over-lapping river. The overlapping river is fixed.

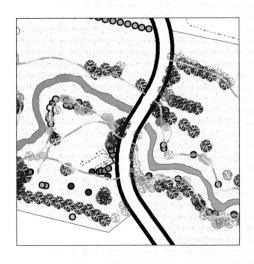

5   Load and zoom to the Lake at Dog Park bookmark. Use the Select tool on the Representation toolbar and click the lake. Notice that when it becomes the selected item, the nodes for the linework are displayed.

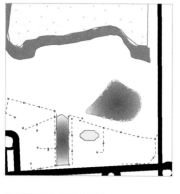

6   Click the Direct Select tool on the Representation toolbar.

7   Use the Direct Select tool to draw a selection box around one of the nodes along the upper part of the lake. The selected node will have a solid fill, while the other nodes are white. Drag the node up close to the river. Doing this will reshape the lake.

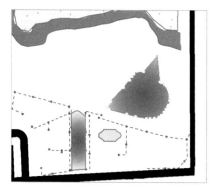

YOUR TURN

One by one, drag the nodes of the lake closer to the river to enlarge the feature and fill in the empty area. When you are done, clear the selected features to examine the results.

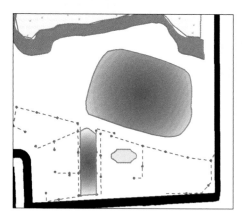

9–1

9–2

The new lake representation is very different from the original feature. Remember that these changes are stored as an override for this one representation only and do not affect the rules you defined for the other water features.

8 Save your map document as **\GIST3\MyAnswers\Tutorial 9-1.mxd**. If you are not continuing to the next exercise, exit ArcMap.

Exercise 9–1

The tutorial demonstrated the process of building and editing cartographic representations.

In this exercise, you will repeat the process to make a sample map for another part of town.

- Start ArcMap and open Tutorial 9–1.mxd.

- Load and zoom to the Map 2 bookmark.

- If you continued from the previous tutorial, turn off the Representation Group.

- Create a new group layer and call it **Representation Group 2**.

- Create new representations for the current map extent from the feature classes in the Raw Data Group. Move them into the new Representation Group 2.

- Build representation rules for all the new layers as you did in tutorial 9–1.

- Save your results as **\GIST3\MyAnswers\Exercise 9-1.mxd**.

WHAT TO TURN IN

If you are working in a classroom setting with an instructor, you may be required to submit the maps you created in tutorial 9–1.

Printed 11-by-17-inch map or screen capture of

       Tutorial 9–1

       Exercise 9–1

Tutorial 9-1 review

Cartographic representations are a new method for adding a rich cartographic style to your maps. Features can be represented by multilayered symbols that can be manipulated by the user.

Rules are made for each layer, and all the features are symbolized according to these rules. But users can also make custom symbols for individual features, allowing them to deviate from the rules. Although it can be time consuming to create these custom symbols, the capabilities of cartographic representations are limitless.

STUDY QUESTIONS

1. Name three ways that cartographic representations are better than the standard symbols.
2. What is the link between cartographic representations and features, and why is maintaining that link important?
3. What are the types of rules that can be built with cartographic representations?

Other real-world examples

Maps are commonly made for tourists to use at national parks such as the Grand Canyon, Yosemite, and Yellowstone. These maps often use physical features of the landscape as points of reference. Making these features look as realistic as possible makes the maps easier to read and interpret by general users.

Historic map reproductions use cartographic representations to mimic a specific style of cartography. Custom symbols can be layered in to make antique-style edging on lakes and oceans, or make unique styles to depict mountains and valleys.

USGS quad sheets have a unique style that makes them useful for hiking. The USGS has created a rich set of custom symbols that can be easily interpreted by hikers who don't have a lot of cartographic knowledge. This adds an intuitive aspect to these maps, which can sometimes be difficult to achieve.

9–1

9–2

Tutorial 9–2

Creating a custom legend

Legends are the devices used on a map to bridge the gap between its symbology and the reality it is depicting. The legend may identify features or the results of an analysis, and are limited only by the tools available in the software. ArcGIS 10 includes many more legend options than previous versions to let cartographers have closer control of the look and style of their maps. As with cartographic representations, the new legend tools can be used to mimic many of the hand-drawing techniques from pre-digital days.

Learning objectives
- *Customize legend shapes*
- *Set up legend fonts*
- *Set up column options*
- *Customize legend elements*

Introduction

One drawback of digital mapping has always been the limitations of legends. There were few options to control the look and style of the legend, and consequently, digital maps often had a mechanical look.

The new legend tools in ArcGIS 10 give much more control of the legend to users. One of these enhancements is the ability to use any font on any part of the legend. Each data layer that is depicted in the legend can have a unique font, perhaps even one that will match the style of the data. The spacing of the fonts can also be controlled.

Another important enhancement is control over the number of columns that a legend uses. The data layers can be grouped into columns, or appear in a column of their own. This can keep a set of legend items from starting in one column and overflowing into another. Instead, the layers can be kept together to lend continuity to the legend style.

One of the best enhancements in ArcGIS 10 is the ability to define the shape of the patches shown in legends. Previously, the only choice for lines was a three-segment zigzag, and the choice for polygons was a rectangle. The new options allow the user to define a custom shape for a patch, whether it be a specialty shape, a feature on the map, or a user-created graphic.

There is an old-school method of creating stylish legends. It involves making a simple legend, converting it to a graphic, and editing the graphic in another program. Then, the legend is added back into the map as a graphic image. The problem is that the link between the table of contents and the legend is lost. Changes made to the way the layers are displayed in the table of contents will not be reflected in the legend.

By combining the new options, the user can create a legend style that fits the data. For example, the patch for a lake can be shaped like a lake, or the linear symbol for a pipe system can be more angular to simulate the way pipe joints really look.

Customizing the legend

Scenario   The Oleander City Council needs some information about library patronage, and you have prepared a map for them to use. It includes point, line, and polygon features, but the standard legend looks a little bland. You'll use some new legend techniques to make it more impressive.

Data   The map includes graduated symbols showing library patronage. Also included are the creeks and bodies of water. The street centerlines and property ownership are added for background interest, and the boundaries of the apartment complexes will be used to compare patronage to multifamily dwelling locations.

Tools used   ArcMap:

Legend Editor

New Legend Patch Shape

A legend has been added to the map. It needs to be placed at the bottom of the layout, but in its current format, it won't fit. First, you'll fix the fonts, then organize the legend into columns, and finally, choose patch shapes that more closely represent the data.

Alter the legend fonts

1 **Start ArcMap and open Tutorial 9–2.mxd. This is a view of the map and the legend.** Some of the layers have cryptic file names, which are not a presentable name for a map. The name that appears in the legend should *never* have cryptic codes, underscores, file extensions (such as .shp), or anything else that is not normal text.

9-1

9-2

2 Click PatronLocation_Aggregated in the table of contents and replace it with **Library Patron Locations**. Conversely, you could open the layer's properties and change the name by using the General tab. Notice that the change made in the table of contents is duplicated automatically in the legend. As you go through and update all the layer names, watch how they appear in the legend.

YOUR TURN

Change the names of the following layers as shown:

Street_Centerlines to **Streets**

ApartmentComplexes to **Apartment Complexes**

Property Ownership to **Oleander Property**

Verify the changes in the legend.

---

3 Next, fix the layer descriptions. Under the Streets layer, change the description CLASS to **Road Classification**. Then change the description of the Oleander Property layer to **Plat Status**. Besides changing the layer name and description, you are also able to change the description of the layer classification directly in the table of contents. This is much faster than opening the properties and changing them in the symbology editor.

4 To add visual interest and distinction to the layer descriptions, you will change their font to italic. Right-click the legend, and then click Properties. Next click the Items tab. In the Legend Items column, click the Library Patron Locations layer. In the Change Text area, click "Selected item(s)." Then click the drop-down arrow and select "Apply to the heading." From the drop-down menu, you can see that each component of the layer can be altered independently.

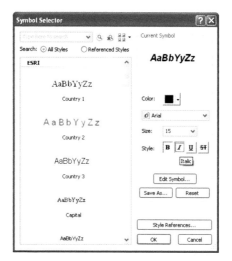

5 Now click Symbol. In the Current Symbol pane, click the Italic button to change the font to italic. Click OK to accept the change.

6 Click Apply and note the changes in the legend. The heading "Number of Visits" is now italic.

Each line in the legend has a specific name that can be changed in the Properties dialog box. The accompanying graphic demonstrates these names.

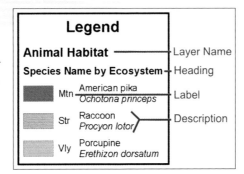

YOUR TURN

Change the headings for the Streets layer and the Oleander Property layer to italic. Click Apply to see the changes, but leave Legend Properties open.

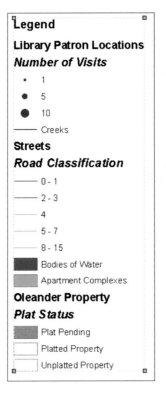

Arrange the legend elements

This has improved the look of the fonts used. Now you will move the layers so that similar layers are grouped together.

1 Select the Bodies of Water layer and click the Up arrow to move Bodies of Water below Creeks. Click Apply.

2 Next, select the Apartment Complexes layer and move it up the list under Library Patron Locations. Click Apply This has changed the order of the layers in the legend, but it has not altered the order of the layers in either the table of contents or the way in which they are drawn on the map.

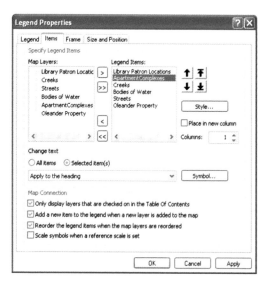

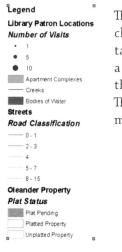

The legend needs to be changed into a horizontal format rather than a vertical one to match the map page layout. This is done by adding more columns.

3 Select the layers Creeks, Streets, and Oleander Property. Select the "Place in new column" check box (**Hint:** Press and hold the Ctrl key while selecting the layers). This will start a new column before each of these layers. Click OK to close the dialog box.

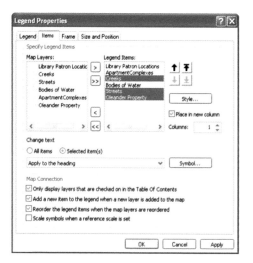

4 Move the legend to the lower part of the map sheet to see the results of the column settings.

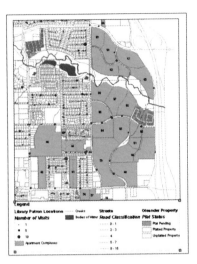

5 The Creeks layer needs more explanation to describe where the data comes from. Open the properties of the Creeks layer, and then click the Symbology tab. In the Legend pane, change the label to **Creeks and Rivers**. This will change the label without changing the layer name. Then click Description and type **Centerline data derived from**. To go to the next line, press Ctrl + M. Then type **2009 aerial photos**. Click OK, and then OK again to close Layer Properties.

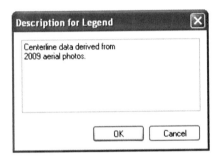

6 The descriptions help, but the symbols used are not indicative of creeks or water. Open Legend Properties, click the Items tab, and then select the Creeks layer. Click Style, and then click Properties. On the General tab, select the "Override default patch" check box. Next click the Line drop-down arrow, and then click Flowing Water. This will change the shape of the patch used for the Creeks layer only.

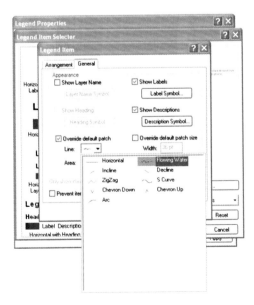

7   Click OK three times to close all the dialog boxes and view the results.

YOUR TURN

Change the patch for Bodies of Water to the Water Body symbol. (**Hint:** Click the Area drop-down arrow.)

Create a custom legend patch

The legend is starting to take on a custom look. The rectangles representing the property lines look wrong, so next you'll make a custom shape for the Oleander Property layer. This will require you to add a new tool called **New Legend Patch Shape** to the Layout menu located in the Page Layout category. (If you do not remember how to add a tool to a toolbar, review tutorial 6–1.)

9–1

9–2

1 Add the New Legend Patch Shape tool to the Layout menu. Set Oleander Property as the only selectable layer and then select a representative sample that includes a few parcels. The outline of what you choose will become the patch symbol.

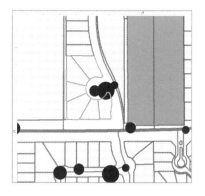

2 Click the New Legend Patch Shape tool. In the Patch Shape pane, click Outline of Features, select the Oleander Property layer, and in the Features drop-down list, click "selected." Then click Create Patch to see a preview of the new patch.

3 Click Add to Styleset and specify a name of **Oleander Property**. Click OK, and then click Close.

4 The patch is stored in your personal styleset. Verify this by clicking Customize on the main menu, and then clicking Style Manager.

5 Expand your personal settings folder and click Area Patches. You'll see your custom patch shape. Close Style Manager.

6 Open Legend Properties again and click the Items tab. Select Oleander Property, click Style, and then click Properties to return to the custom patch settings. Make sure you are on the General tab. Select the "Override default patch" check box and review the Area drop-down list. You'll see your custom patch. Click it, and then close all the dialog boxes. The display changes to show your custom patch, shaded in the Symbology colors. Notice that although these changed in the legend, they did not change in the table of contents.

7 The Apartment Complexes layer is opaque and covers the Oleander Property data. To fix this, open the properties of the layer, click the Display tab, and set the transparency to 70%. Click OK to close the Layer Properties dialog box. There is a problem, though. Note that the color in the legend does not match the color in the map.

8 To fix this transparency problem, open the properties of the Patron Analysis data frame. Click the General tab and select the "Simulate layer transparency in legends" check box.

9 Close Properties and note the change in the legend. Now the colors match.

10 Finally, increase the spacing between columns. Open Legend Properties and click the Legend tab. In the Spacing between list, click columns and increase the size to **14**. Note the other controls that are available, and then click OK.

11 You can resize the legend to fit in your map display by selecting one of the corners and dragging it inward to make the box smaller. Then move the legend into the lower half of the map. This legend, with its own unique personal touches, looks much better than a standard, out-of-the-box legend. It makes the statement that you are as concerned about how your map is perceived as you are about the data that it displays.

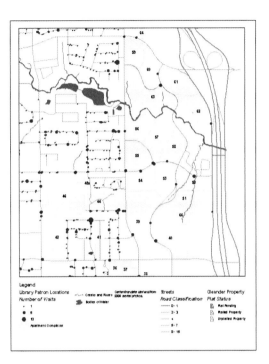

12 Save your map document as **\GIST3\MyAnswers\Tutorial 9-2.mxd**. If you are not continuing to the next exercise, exit ArcMap.

Exercise 9–2

The tutorial showed techniques for altering the standard legend and adding custom touches.

In this exercise, you will try altering another legend using the same techniques.

- Start ArcMap and open your results from Exercise 9–1.mxd.
- Load and zoom to the Exercise 9–2 bookmark.
- Change the names of the layers to better reflect their contents.
- Change the fonts for the titles, descriptions, and so forth, as you see fit.
- Alter the number of columns to best display the data.
- Build a custom patch shape and apply it to one of the layers.
- Set the spacing as needed and move the legend to an appropriate place on the map.
- Save your results as **\GIST3\MyAnswers\Exercise 9–2.mxd**.

WHAT TO TURN IN

If you are working in a classroom setting with an instructor, you may be required to submit the maps you created in tutorial 9-2.

Printed 11-by-17-inch map or screen capture of

Tutorial 9-2

Exercise 9-2

Tutorial 9–2 review

The realism of a legend can make or break your map. If a viewer can't interpret the results and recognize features by their symbology, they will have a hard time appreciating the analysis that you may have done. Cartographic representations give you the tools to make your map look more realistic. Carry on that theme by using the legend tools to make your legends just as realistic.

The tools include changing fonts and spacing for all the text in the legend. Any font that can be loaded into your operating system font manager can be used in the legend. The customization continues with the ability to set the number of columns. This can help group your layers into similar categories or symbol styles. Finally, the new tools give you the ability to create custom shapes for the legend patches. This can add a custom touch that will be unique to your map.

Investigate each of the options for the font, spacing, columns, and patches for each layer to make your legend distinct and stylish.

STUDY QUESTIONS

1. How does the style of font affect the way viewers interpret your map?

2. What data types might be a good choice for a custom patch shape?

3. How would you use columns to emphasize a set of data?

Other real-world examples

A service company that makes replicas of historic maps has the task of trying to reproduce the elegance of hand-drawn legends. The legends use various text styles, shapes, and multiple columns to create their unique look. Using the ArcGIS legend tools and techniques, it is possible to make a legend that will approach the style of the hand-drawn legend.

Maps that show physical feature data, such as USGS quad sheets, need to have symbols that more closely match the data they are representing. By using custom legend options, these maps can have a very natural look and feel.

References for further study

There are a number of additional resources to help develop your cartographic skills. **ArcGIS Desktop Help** lets you use keywords to search for a title or topic. You can access ArcGIS Desktop Help in ArcMap or ArcCatalog by clicking Help on the main menu. Then click the Search tab, type a keyword, and click Ask. Use the keywords provided here to search for additional learning resources on many of the concepts taught in this chapter.

ESRI has developed a number of online courses on a wide variety of pertinent GIS topics. **ESRI Self-Study (Virtual Campus) courses** are excellent resources for students that will supplement information covered in this course. ESRI Virtual Campus courses can be accessed at `http://training.esri.com/gateway/index.cfm`.

ArcGIS Desktop Help search keywords

cartographic representations, symbology, symbol layers, markers, maps, cartography, legends, map templates, styles, cartography tools, layer properties, graphic quality toolset

ESRI Self-Study (Virtual Campus) courses

The following ESRI Virtual Campus classes may be helpful:

1. Cartographic Design Using ArcGIS 9
2. Introduction to Cartographic Representations in ArcGIS 9.2
3. Making Better Map Layouts with ArcGIS
4. What's New in ArcGIS 9 Labeling and Annotation

Appendix A

Task index

Appendix B

Data source credits

Chapter 1 data sources include
All data created new by student using geodatabase forms designed by the author

Chapter 2 data sources include
All data created new by student using geodatabase forms designed by the author

Chapter 3 data sources include
\ESRIPress\GIST3\Data\ParcelLineSource.shp, derived from City of Euless
\ESRIPress\GIST3\Data\ParcelSource.shp, derived from City of Euless
\ESRIPress\GIST3\Data\ZoningLineSource.shp, derived from City of Euless
\ESRIPress\GIST3\Data\ZoningSource.shp, derived from City of Euless
\ESRIPress\GIST3\Data\SewerLineSource.shp, derived from City of Euless
\ESRIPress\GIST3\Data\SewerNodeSource.shp, derived from City of Euless
\ESRIPress\GIST3\Data\Storm_Fix_Source.shp, derived from City of Euless
\ESRIPress\GIST3\Data\Storm_Line_Source.shp, derived from City of Euless

Chapter 4 data sources include
\ESRIPress\GIST3\Data\monument.shp, derived from City of Euless
\ESRIPress\GIST3\Data\ORTA.mdb\ProposedFacilities\ConstructionLinear, created by author
\ESRIPress\GIST3\Data\ORTA.mdb\ProposedFacilities\ConstructionPolygons, created by author
\ESRIPress\GIST3\Data\ORTA.mdb\ProposedFacilities\RailLines, created by author
\ESRIPress\GIST3\Data\CityOfOleander.mdb\Property Data\LotBoundaries, derived from City of Euless
\ESRIPress\GIST3\Data\CityOfOleander.mdb\Property Data\Parcels, derived from City of Euless

\ESRIPress\GIST3\Data\monument.shp, derived from City of Euless
\ESRIPress\GIST3\Data\ORTA.mdb\ProposedFacilities\ConstructionLinear, created by author
\ESRIPress\GIST3\Data\ORTA.mdb\ProposedFacilities\ConstructionPolygons, created by author
\ESRIPress\GIST3\Data\ORTA.mdb\ProposedFacilities\RailLines, created by author
\ESRIPress\GIST3\Data\CityOfOleander.mdb\Property Data\LotBoundaries, derived from City of Euless
\ESRIPress\GIST3\Data\CityOfOleander.mdb\Property Data\Parcels, derived from City of Euless
\ESRIPress\GIST3\Data\monument.shp, derived from City of Euless
\ESRIPress\GIST3\Data\ORTA.mdb\ProposedFacilities\ConstructionLinear, created by author
\ESRIPress\GIST3\Data\ORTA.mdb\ProposedFacilities\ConstructionPolygons, created by author
\ESRIPress\GIST3\Data\ORTA.mdb\ProposedFacilities\RailLines, created by author
\ESRIPress\GIST3\Data\CityOfOleander.mdb\Property Data\LotBoundaries, derived from City of Euless
\ESRIPress\GIST3\Data\CityOfOleander.mdb\Property Data\Parcels, derived from City of Euless
\ESRIPress\GIST3\Data\monument.shp, derived from City of Euless
\ESRIPress\GIST3\Data\CityOfOleander.mdb\Property Data\LotBoundaries, derived from City of Euless
\ESRIPress\GIST3\Data\monument.shp, derived from City of Euless
\ESRIPress\GIST3\Data\CityOfOleander.mdb\Property Data\LotBoundaries, derived from City of Euless
\ESRIPress\GIST3\Data\Images\The Landing - Plat Segment.tif, courtesy of Thomas Hoover
\ESRIPress\GIST3\Data\Images\Bear Creek Estates IV.tif, courtesy of Steve Miller
\ESRIPress\GIST3\Data\Images\Creekwood Estates.tif, courtesy of Steve Miller
\ESRIPress\GIST3\Data\Images\Little Bear II.tif, courtesy of Steve Miller
\ESRIPress\GIST3\Data\Images\Running Bear Estates.tif, courtesy of Steve Miller

Chapter 5 data sources include
\ESRIPress\GIST3\Data\CityOfOleander.mdb\PoliceData\PoliceDistricts, derived from City of Euless
\ESRIPress\GIST3\Data\CityOfOleander.mdb\PoliceData\PoliceDistricts_Boundaries, derived from City of Euless
\ESRIPress\GIST3\Data\CityOfOleander.mdb\City_Limit, derived from City of Euless
\ESRIPress\GIST3\Data\Geometric Networks.mdb\SewerNetwork\SewerConnectors, derived from City of Euless
\ESRIPress\GIST3\Data\Geometric Networks.mdb\SewerNetwork\SewerLines, derived from City of Euless
\ESRIPress\GIST3\Data\Geometric Networks.mdb\Symbology\SewerConnectorSymb, derived from City of Euless
\ESRIPress\GIST3\Data\Geometric Networks.mdb\Symbology\SewerLinesSymb, derived from City of Euless
\ESRIPress\GIST3\Data\Geometric Networks.mdb\WaterDistribution\DistLateral, derived from City of Euless
\ESRIPress\GIST3\Data\Geometric Networks.mdb\WaterDistribution\DistMains, derived from City of Euless
\ESRIPress\GIST3\Data\Geometric Networks.mdb\WaterDistribution\Fittings, derived from City of Euless
\ESRIPress\GIST3\Data\Geometric Networks.mdb\WaterDistribution\HydLaterals, derived from City of Euless
\ESRIPress\GIST3\Data\Geometric Networks.mdb\WaterDistribution\WaterDistribution_Net, derived from City of Euless
\ESRIPress\GIST3\Data\Geometric Networks.mdb\WaterDistribution\WaterDistribution_Net_Junctions, derived from City of Euless
\ESRIPress\GIST3\Data\CityOfOleander.mdb\SampleData\TheLandingPolygons, derived from City of Euless
\ESRIPress\GIST3\Data\Images\GeometricNetworkCompleteRectify.tif, courtesy of Thomas Hoover
\ESRIPress\GIST3\Data\CityOfOleander.mdb\Property Data\Blocks, derived from City of Euless
\ESRIPress\GIST3\Data\CityOfOleander.mdb\Property Data\CityArea, derived from City of Euless
\ESRIPress\GIST3\Data\CityOfOleander.mdb\Property Data\CityBoundary, derived from City of Euless
\ESRIPress\GIST3\Data\CityOfOleander.mdb\Property Data\LandingPerimeter, derived from City of Euless
\ESRIPress\GIST3\Data\CityOfOleander.mdb\Property Data\LotBoundaries, derived from City of Euless
\ESRIPress\GIST3\Data\CityOfOleander.mdb\Property Data\Parcels, derived from City of Euless

Chapter 6 data sources include

\ESRIPress\GIST3\Data\CityOfOleander.mdb\Property Data\CityBoundary, derived from City of Euless

\ESRIPress\GIST3\Data\CityOfOleander.mdb\Property Data\LotBoundaries, derived from City of Euless

\ESRIPress\GIST3\Data\CityOfOleander.mdb\PoliceData\PoliceDistricts, derived from City of Euless

\ESRIPress\GIST3\Data\CityOfOleander.mdb\PoliceData\PoliceDistricts_Boundaries, derived from City of Euless

\ESRIPress\GIST3\Data\CityOfOleander.mdb\LibraryData\PatronLocation, derived from City of Euless

\ESRIPress\GIST3\Data\CityOfOleander.mdb\StreetData\Street_Centerlines, derived from City of Euless

\ESRIPress\GIST3\Data\Images\The Landing - Plat.tif, courtesy of Thomas Hoover

Chapter 7 data sources include

\ESRIPress\GIST3\Data\CityOfOleander.mdb\LibraryData\ApartmentComplexes, derived from City of Euless

\ESRIPress\GIST3\Data\CityOfOleander.mdb\LibraryData\PatronLocation_Aggregated, derived from City of Euless

\ESRIPress\GIST3\Data\CityOfOleander.mdb\StreetData\Street_Centerlines, derived from City of Euless

\ESRIPress\GIST3\Data\CityOfOleander.mdb\Property Data\Parcels, derived from City of Euless

\ESRIPress\GIST3\Data\Apartments.txt, derived from City of Euless

\ESRIPress\GIST3\Data\CityOfOleander.mdb\LibraryData\ApartmentComplexes, derived from City of Euless

\ESRIPress\GIST3\Data\CityOfOleander.mdb\LibraryData\PatronLocation_Aggregated, derived from City of Euless

\ESRIPress\GIST3\Data\CityOfOleander.mdb\StreetData\Street_Centerlines, derived from City of Euless

\ESRIPress\GIST3\Data\CityOfOleander.mdb\Property Data\Parcels, derived from City of Euless

\ESRIPress\GIST3\Data\Apartments.txt, derived from City of Euless

\ESRIPress\GIST3\Data\CityOfOleander.mdb\LibraryData\ApartmentComplexes, derived from City of Euless

\ESRIPress\GIST3\Data\CityOfOleander.mdb\LibraryData\PatronLocation_Aggregated, derived from City of Euless

\ESRIPress\GIST3\Data\CityOfOleander.mdb\StreetData\Street_Centerlines, derived from City of Euless

\ESRIPress\GIST3\Data\CityOfOleander.mdb\Property Data\Parcels, derived from City of Euless

\ESRIPress\GIST3\Data\Apartments.txt, derived from City of Euless

Chapter 8 data sources include

\ESRIPress\GIST3\Data\SewerMaps.mdb\SewerMaps\sewerlin, derived from City of Euless

\ESRIPress\GIST3\Data\SewerMaps.mdb\SewerMaps\sewernod, derived from City of Euless

\ESRIPress\GIST3\Data\CityOfOleander.mdb\Property Data\Blocks, derived from City of Euless

\ESRIPress\GIST3\Data\CityOfOleander.mdb\Property Data\LotBoundaries, derived from City of Euless

\ESRIPress\GIST3\Data\CityOfOleander.mdb\Property Data\Parcels, derived from City of Euless

\ESRIPress\GIST3\Data\CityOfOleander.mdb\Property Data\Plat_Index, derived from City of Euless

\ESRIPress\GIST3\Data\CityOfOleander.mdb\Property Data\ZoningDistricts, derived from City of Euless

\ESRIPress\GIST3\Data\CityOfOleander.mdb\StreetData\Street_Centerlines, derived from City of Euless

\ESRIPress\GIST3\Data\CityOfOleander.mdb\Property Data\LotBoundaries, derived from City of Euless

\ESRIPress\GIST3\Data\CityOfOleander.mdb\Property Data\ZoningBoundaries, derived from City of Euless

\ESRIPress\GIST3\Data\CityOfOleander.mdb\Property Data\ZoningDistricts, derived from City of Euless

\ESRIPress\GIST3\Data\CityOfOleander.mdb\Property Data\ZoningDistrictsMasks, derived from City of Euless

\ESRIPress\GIST3\Data\CityOfOleander.mdb\StreetData\Street_Centerlines, derived from City of Euless

Chapter 9 data sources include

\ESRIPress\GIST3\Data\Planimetrics.mdb\Data_2009\BodiesOfWater, derived from City of Euless

\ESRIPress\GIST3\Data\Planimetrics.mdb\Data_2009\Creeks, derived from City of Euless

\ESRIPress\GIST3\Data\Planimetrics.mdb\Data_2009\MiscStructures, derived from City of Euless

\ESRIPress\GIST3\Data\Planimetrics.mdb\Data_2009\Recreation, derived from City of Euless

\ESRIPress\GIST3\Data\Planimetrics.mdb\Data_2009\Street_Centerlines, derived from City of Euless

\ESRIPress\GIST3\Data\Planimetrics.mdb\Data_2009\Trails, derived from City of Euless

\ESRIPress\GIST3\Data\Planimetrics.mdb\Data_2009\TreeInventory_Midway, derived from City of Euless

\ESRIPress\GIST3\Data\Planimetrics.mdb\Data_2009\TreeInventory_Oleander, derived from City of Euless

\ESRIPress\GIST3\Data\CityOfOleander.mdb\LibraryData\ApartmentComplexes, derived from City of Euless

\ESRIPress\GIST3\Data\CityOfOleander.mdb\LibraryData\PatronLocation, derived from City of Euless

\ESRIPress\GIST3\Data\CityOfOleander.mdb\LibraryData\PatronLocation_Aggregated, derived from City of Euless

\ESRIPress\GIST3\Data\CityOfOleander.mdb\StreetData\Street_Centerlines, derived from City of Euless

\ESRIPress\GIST3\Data\CityOfOleander.mdb\Property Data\Parcels, derived from City of Euless

\ESRIPress\GIST3\Data\Planimetrics.mdb\Data_2009\BodiesOfWater, derived from City of Euless

\ESRIPress\GIST3\Data\Planimetrics.mdb\Data_2009\Creeks, derived from City of Euless

\ESRIPress\GIST3\Data\Apartments.txt, derived from City of Euless

\ESRIPress\GIST3\Data\Planimetrics.mdb\Data_2009\MiscStructures, derived from City of Euless

\ESRIPress\GIST3\Data\Planimetrics.mdb\Data_2009\Recreation, derived from City of Euless

\ESRIPress\GIST3\Data\Planimetrics.mdb\Data_2009\Street_Centerlines, derived from City of Euless

\ESRIPress\GIST3\Data\Planimetrics.mdb\Data_2009\Trails, derived from City of Euless

\ESRIPress\GIST3\Data\Planimetrics.mdb\Data_2009\TreeInventory_Midway, derived from City of Euless

\ESRIPress\GIST3\Data\Planimetrics.mdb\Data_2009\TreeInventory_Oleander, derived from City of Euless

Appendix C

Data license agreement

Important:
Read carefully before opening the sealed media package

ENVIRONMENTAL SYSTEMS RESEARCH INSTITUTE INC. (ESRI) IS WILLING TO LICENSE THE ENCLOSED DATA AND RELATED MATERIALS TO YOU ONLY UPON THE CONDITION THAT YOU ACCEPT ALL OF THE TERMS AND CONDITIONS CONTAINED IN THIS LICENSE AGREEMENT. PLEASE READ THE TERMS AND CONDITIONS CARE-FULLY BEFORE OPENING THE SEALED MEDIA PACKAGE. BY OPENING THE SEALED MEDIA PACKAGE, YOU ARE INDICATING YOUR ACCEPTANCE OF THE ESRI LICENSE AGREEMENT. IF YOU DO NOT AGREE TO THE TERMS AND CONDITIONS AS STATED, THEN ESRI IS UNWILLING TO LICENSE THE DATA AND RELATED MATERIALS TO YOU. IN SUCH EVENT, YOU SHOULD RETURN THE MEDIA PACKAGE WITH THE SEAL UNBROKEN AND ALL OTHER COMPONENTS TO ESRI.

ESRI license agreement

This is a license agreement, and not an agreement for sale, between you (Licensee) and Environmental Systems Research Institute Inc. (ESRI). This ESRI License Agreement (Agreement) gives Licensee certain limited rights to use the data and related materials (Data and Related Materials). All rights not specifically granted in this Agreement are reserved to ESRI and its Licensors.

Reservation of Ownership and Grant of License: ESRI and its Licensors retain exclusive rights, title, and ownership to the copy of the Data and Related Materials licensed under this Agreement and, hereby, grant to Licensee a personal, nonexclusive, nontransferable, royalty-free, worldwide license to use the Data and Related Materials based on the terms and conditions of this Agreement. Licensee agrees to use reasonable effort to protect the Data and Related Materials from unauthorized use, reproduction, distribution, or publication.

Proprietary Rights and Copyright: Licensee acknowledges that the Data and Related Materials are proprietary and confidential property of ESRI and its Licensors and are protected by United States copyright laws and applicable international copyright treaties and/or conventions.

Permitted Uses: Licensee may install the Data and Related Materials onto permanent storage device(s) for Licensee's own internal use.

Licensee may make only one (1) copy of the original Data and Related Materials for archival purposes during the term of this Agreement unless the right to make additional copies is granted to Licensee in writing by ESRI.

Licensee may internally use the Data and Related Materials provided by ESRI for the stated purpose of GIS training and education.

Uses Not Permitted: Licensee shall not sell, rent, lease, sublicense, lend, assign, time-share, or transfer, in whole or in part, or provide unlicensed Third Parties access to the Data and Related Materials or portions of the Data and Related Materials, any updates, or Licensee's rights under this Agreement.

Licensee shall not remove or obscure any copyright or trademark notices of ESRI or its Licensors.

Term and Termination: The license granted to Licensee by this Agreement shall commence upon the acceptance of this Agreement and shall continue until such time that Licensee elects in writing to discontinue use of the Data or Related Materials and terminates this Agreement. The Agreement shall automatically terminate without notice if Licensee fails to comply with any provision of this Agreement. Licensee shall then return to ESRI the Data and Related Materials. The parties hereby agree that all provisions that operate to protect the rights of ESRI and its Licensors shall remain in force should breach occur.

Disclaimer of Warranty: The Data and Related Materials contained herein are provided "as-is," without warranty of any kind, either express or implied, including, but not limited to, the implied warranties of merchantability, fitness for a particular purpose, or noninfringement. ESRI does not warrant that the Data and Related Materials will meet Licensee's needs or expectations, that the use of the Data and Related Materials will be uninterrupted, or that all nonconformities, defects, or errors can or will be corrected. ESRI is not inviting reliance on the Data or Related Materials for commercial planning or analysis purposes, and Licensee should always check actual data.

Data Disclaimer: The Data used herein has been derived from actual spatial or tabular information. In some cases, ESRI has manipulated and applied certain assumptions, analyses, and opinions to the Data solely for educational training purposes. Assumptions, analyses, opinions applied, and actual outcomes may vary. Again, ESRI is not inviting reliance on this Data, and the Licensee should always verify actual Data and exercise their own professional judgment when interpreting any outcomes.

LIMITATION OF LIABILITY: ESRI SHALL NOT BE LIABLE FOR DIRECT, INDIRECT, SPECIAL, INCIDENTAL, OR CONSEQUENTIAL DAMAGES RELATED TO LICENSEE'S USE OF THE DATA AND RELATED MATERIALS, EVEN IF ESRI IS ADVISED OF THE POSSIBILITY OF SUCH DAMAGE.

No Implied Waivers: No failure or delay by ESRI or its Licensors in enforcing any right or remedy under this Agreement shall be construed as a waiver of any future or other exercise of such right or remedy by ESRI or its Licensors.

Order for Precedence: Any conflict between the terms of this Agreement and any FAR, DFAR, purchase order, or other terms shall be resolved in favor of the terms expressed in this Agreement, subject to the government's minimum rights unless agreed otherwise.

Export Regulation: Licensee acknowledges that this Agreement and the performance thereof are subject to compliance with any and all applicable United States laws, regulations, or orders relating to the export of data thereto. Licensee agrees to comply with all laws, regulations, and orders of the United States in regard to any export of such technical data.

Severability: If any provision(s) of this Agreement shall be held to be invalid, illegal, or unenforceable by a court or other tribunal of competent jurisdiction, the validity, legality, and enforceability of the remaining provisions shall not in any way be affected or impaired thereby.

Governing Law: This Agreement, entered into in the County of San Bernardino, shall be construed and enforced in accordance with and be governed by the laws of the United States of America and the State of California without reference to conflict of laws principles. The parties hereby consent to the personal jurisdiction of the courts of this county and waive their rights to change venue.

Entire Agreement: The parties agree that this Agreement constitutes the sole and entire agreement of the parties as to the matter set forth herein and supersedes any previous agreements, understandings, and arrangements between the parties relating hereto.

Appendix D

Installing the data and software

GIS Tutorial 3: Advanced Workbook includes a DVD containing exercise data and a DVD containing a fully functioning 180-day trial version of ArcGIS Desktop 10 software (ArcEditor license level). If you already have a licensed copy of ArcGIS Desktop 10 installed on your computer (or have access to the software through a network), do not install the trial software. Use your licensed software to do the exercises in this book. If you have an older version of ArcGIS installed on your computer, you must uninstall it before you can install the software that is provided with this book.

.NET Framework 3.5 SP1 must be installed on your computer before you install ArcGIS Desktop 10. Some features of ArcGIS Desktop 10 require Microsoft Internet Explorer version 8.0. If you do not have Microsoft Internet Explorer version 8.0, you must install it before installing ArcGIS Desktop 10.

Installing the exercise data

Follow the steps below to install the exercise data.

1 Put the data DVD in your computer's DVD drive. A splash screen will appear.

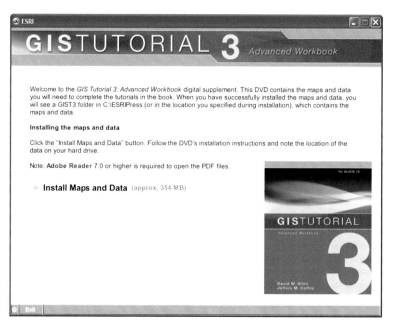

2 Read the welcome, then click the Install Exercise Data link. This launches the InstallShield Wizard.

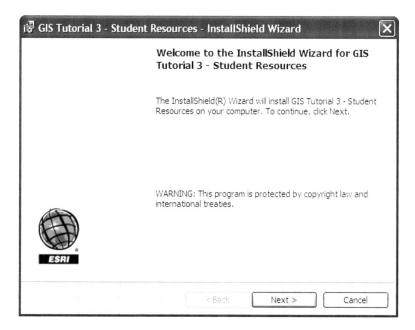

3 Click Next. Read and accept the license agreement terms, then click Next.

4   Accept the default installation folder or click Browse and navigate to the drive or folder location where you want to install the data.

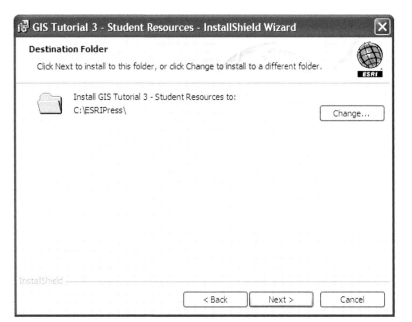

5   Click Next. The installation will take a few moments. When the installation is complete, you will see the following message.

6   Click Finish. The exercise data is installed on your computer in a folder called C:\ESRIPress\GIST3.

Uninstalling the data and resources

To uninstall the data and resources from your computer, open your operating system's control panel and double-click the Add/Remove Programs icon. In the Add/Remove Programs dialog box, select the following entry and follow the prompts to remove it:

GIS Tutorial 3

Installing the software

The ArcGIS software included with this book is intended for educational purposes only. Once installed and registered, the software will run for 180 days. The software cannot be reinstalled nor can the time limit be extended. It is recommended that you uninstall this software when it expires.

> ### IMPORTANT
> Before installing the trial software, visit www.esri.com/180daytrial to activate your authorization code. Visit www.esri.com/evalhelp for evaluation software support.

Follow the steps below to install the software.

1 Put the software DVD in your computer's DVD drive. A splash screen will appear. If your auto-run is disabled, navigate to the contents of the DVD and double-click the ESRI.exe file to begin.

2 Click the ArcGIS Desktop Setup installation option. This will launch the Setup wizard.

3 Read the Welcome screen, and then click Next.

4 Read the license agreement. Click "I accept the license agreement," and then click Next.

5 Choose the Complete install option, which will add extension products that are used in the book. Click Next.

6 Accept the default installation folder or navigate to the drive or folder location where you want to install the software. Click Next.

7 Accept the default installation folder or navigate to the drive or folder where you want to install Python, a scripting language used by some ArcGIS geoprocessing functions. (You won't see this panel if you already have Python installed.) Click Next.

8 The installation paths for ArcGIS and Python are confirmed. Click Next. The software will take some time to install on your computer.

9 Click Finish when the installation is completed.

10 On the ArcGIS Administrator Wizard window, select ArcEditor (Single Use), then click Authorize Now.

11 Select "I have installed my software and need to authorize it." Click Next.

12 Follow the wizard to begin the authorization process. Use the authorization code located at the bottom of the software DVD jacket in the back of the book.

If you have questions or encounter problems during the installation process, or while using this book, please use the resources listed below. (The ESRI Technical Support Department does not answer questions regarding the ArcGIS Desktop software, the GIS Tutorial supplementary media, or the contents of the book itself.)

- To resolve problems with the trial software, visit `www.esri.com/evalhelp`.

- To report problems with the software or exercise data DVDs, or to report mistakes in the book, send an e-mail to ESRI workbook support at `workbook-support@esri.com`.

Uninstalling the software

To uninstall the software from your computer, open your operating system's control panel and double-click the Add/Remove Programs icon. In the Add/Remove Programs dialog box, select the following entry and follow the prompts to remove it:

ArcGIS Desktop 10

Related titles from ESRI Press

Designing Geodatabases: Case Studies in GIS Data Modeling

ISBN: 978-1-58948-021-6

Designing Geodatabases outlines five steps for taking a data model through its conceptual, logical, and physical phases—modeling the user's view, defining objects and relationships, selecting geographic representations, matching geodatabase elements, and organizing the geodatabase structure. Several design models for a variety of applications are considered, including addresses and locations, census units and boundaries, stream and river networks, and topography and the basemap.

Modeling Our World: The ESRI Guide to Geodatabase Concepts, Second Edition

ISBN: 978-1-58948-278-4

Modeling Our World, Second Edition, promotes best practices for data modeling and analysis by addressing critical topics such as spatial integrity, attribute integrity, work flow, and scaling. This book clarifies geographic data modeling concepts and is your complete survey of the geodatabase information model.

GIS, Spatial Analysis, and Modeling

ISBN: 978-1-58948-130-5

Recent advancements in GIS software, along with the availability of spatially referenced data, now makes possible the sophisticated modeling and statistical analysis of all types of geographic phenomena. *GIS, Spatial Analysis, and Modeling* presents a compendium of papers that advance the methods and practices used to develop meaningful spatial analysis for decision support. A resource for geographic analysts, modelers, software engineers, and GIS professionals, this text covers tools, techniques, and methods while providing examples of various applications.

Lining Up Data in ArcGIS: A Guide to Map Projections

ISBN: 978-1-58948-249-4

Lining Up Data in ArcGIS is a practical guide to solving the problem of data misalignment. This book presents techniques to identify data projections and create custom projections to align data. Formatted for practical use, each chapter stands alone, addressing specific issues related to working with coordinate systems. This handbook will benefit beginning and skilled GIS users alike.

ESRI Press publishes books about the science, application, and technology of GIS. Ask for these titles at your local bookstore or order by calling 1-800-447-9778. You can also read book descriptions, read reviews, and shop online at www.esri.com/esripress. Outside the United States, visit our Web site at www.esri.com/esripressorders for a full list of book distributors and their territories.